# Laser–Plasma Accelerators and Radiation Sources

**Online at: https://doi.org/10.1088/978-0-7503-2703-9**

**IOP Series in Coherent Sources, Quantum Fundamentals, and Applications**

**About the Editor**

F J Duarte is a laser physicist based in Western New York, USA. His career has spanned three continents with contributions within the academic, industrial, and defense sectors. Duarte is editor/author of 15 laser optics books and sole author of three books: *Tunable Laser Optics, Quantum Optics for Engineers*, and *Fundamentals of Quantum Entanglement*. Duarte has made original contributions in the fields of coherent imaging, directed energy, high-power tunable lasers, laser metrology, liquid- and solid-state organic gain media, narrow-linewidth tunable laser oscillators, organic semiconductor coherent emission, $N$-slit quantum interferometry, polarization rotation, quantum entanglement, and space-to-space secure interferometric communications. He is also the author of the generalized multiple-prism grating dispersion theory and pioneered the use of Dirac's quantum notation in $N$-slit interferometry and classical optics. His contributions have found applications in numerous fields, including astronomical instrumentation, dispersive optics, femtosecond laser microscopy, geodesics, gravitational lensing, heat transfer, laser isotope separation, laser medicine, laser pulse compression, laser spectroscopy, mathematical transforms, nonlinear optics, polarization optics, and tunable diode-laser design. Duarte was elected Fellow of the Australian Institute of Physics in 1987 and Fellow of the Optical Society of America in 1993. He has received various recognitions, including the *Paul F Foreman Engineering Excellence Award* and the *David Richardson Medal* from the Optical Society.

**Coherent Sources, Quantum Fundamentals, and Applications**

Since its discovery the laser has found innumerable applications from astronomy to zoology. Subsequently, we have also become familiar with additional sources of coherent radiation such as the free electron laser, optical parametric oscillators, and coherent interferometric emitters. The aim of this book Series in Coherent Sources, Quantum Fundamentals, and Applications is to explore and explain the physics and technology of widely applied sources of coherent radiation and to match them with utilitarian and cutting-edge scientific applications. Coherent sources of interest are those that offer advantages in particular emission characteristics areas such as broad tunability, high spectral coherence, high energy, or high power. An additional area of inclusion is the coherent sources capable of high performance in the miniaturized realm. Understanding of quantum fundamentals can lead to new and better coherent sources and unimagined scientific and technological applications. Application areas of interest include the industrial, commercial, and medical sectors. Also, particular attention is given to scientific applications with a bright future such as coherent spectroscopy, astronomy, biophotonics, space communications, space interferometry, quantum entanglement, and quantum interference.

**Publishing benefits**
Authors are encouraged to take advantage of the features made possible by electronic publication to enhance the reader experience through the use of color, animation and video, and incorporating supplementary files in their work.

**Do you have an idea of a book that you'd like to explore?**
For further information and details of submitting book proposals, see iopscience.org/books or contact Ashley Gasque at ashley.gasque@iop.org.

A full list of titles published in this series can be found here: https://iopscience.iop.org/bookListInfo/series-in-coherent-sources-and-applications.

# Laser–Plasma Accelerators and Radiation Sources

**Hyung Taek Kim**
*Advanced Photonics Research Institute, Gwangju Institute of Science and Technology, Gwanju, Republic of Korea*

**IOP** Publishing, Bristol, UK

Multimedia content is available for this book from https://doi.org/10.1088/978-0-7503-2703-9.



ISBN 978-0-7503-2703-9 (ebook)
ISBN 978-0-7503-2701-5 (print)
ISBN 978-0-7503-2704-6 (myPrint)
ISBN 978-0-7503-2702-2 (mobi)

DOI 10.1088/978-0-7503-2703-9

Version: 20231201

IOP ebooks

British Library Cataloguing-in-Publication Data: A catalogue record for this book is available from the British Library.

Published by IOP Publishing, wholly owned by The Institute of Physics, London

IOP Publishing, No.2 The Distillery, Glassfields, Avon Street, Bristol, BS2 0GR, UK

US Office: IOP Publishing, Inc., 190 North Independence Mall West, Suite 601, Philadelphia, PA 19106, USA

*Dedication to my wife and children for putting up with me while I worked on this book.*

# Contents

# Preface

When we have a high-power laser system for laser–plasma interactions, we are in front of two monsters, the laser and the plasma. Once I called our petawatt (PW) laser a monster that is difficult to control at a conference on laser plasma accelerator. Then, Professor C Josh, one of the pioneers of plasma electron accelerator, asked me, 'You mentioned that the petawatt laser is a monster. What will be the upcoming 10 PW or even 100 PW?' I was unable to correctly respond to the question at the time. I had to say, 'The petawatt laser is a monster until you learn how to control it, at which point it will become a pet.' If you are unfamiliar with lasers and plasmas, you may think they are torturing you like living creatures. However, once you learn how to control them, they will provide you unique particle and radiation sources for scientific research. This book is written with that goal in mind. I aim to provide the fundamental and practical knowledge necessary to comprehend and control the high-power laser system and plasma dynamics to produce high-energy particles and radiation sources. In particular, this book will provide practical information on laser–plasma experiments, such as laser control methods, focusing geometry, vacuum chambers, targetry, and detection systems, which are rarely compiled in a single publication.

I have spent over 20 years in laboratories studying laser–plasma interactions with ultra-intense lasers. When I first began my career, I generated higher-order harmonics with a few-terawatt laser. I am currently concentrating on electron acceleration and radiation development with a 4 PW laser. In the past two decades, laser power has increased by orders of magnitude. There are not many research fields that are evolving as rapidly as ultra-intense laser technologies and their interactions with plasmas. However, training a novice scientist to become an expert in laser–plasma interactions is a difficult task. To design, execute, and evaluate laser–plasma experiments, one must be knowledgeable of ultra-intense laser technologies, vacuum techniques, targetries, and detection systems. This includes optics, nonlinear optics, spectroscopy, electromagnetism, quantum electronics, material science, plasma physics, fluid dynamics, and relativistic plasma dynamics, among others. Typically, at least several years are required to study the fundamentals of each of these disciplines. Thus, I would like to transfer my knowledge to novices in this field in order to reduce the amount of time it takes them to comprehend the fundamentals of laser–plasma interactions with ultra-intense lasers. This book will not describe details of fundamental theories because there are many excellent literatures as introduced in the references. This book could possibly be titled 'Ultra-intense Laser–Plasma Experiments for Dummies.' I hope that this book contains the information necessary for designing experimental setups, conducting experiments, and analyzing data.

With best wishes on your adventure with ultra-intense lasers and energetic plasmas.

Hyung Taek Kim

# Acknowledgments

I am grateful to all of my colleagues at the Ultrashort Quantum Beam Facility (UQBF) and the Center for Relativistic Laser Science (CoReLS), as well as collaborators in laser-plasma research, for inspiring and teaching me about laser-plasma physics. Thank you, in particular, to Dr Vishwa Bandhu Pathak for providing the PIC simulation results shown in figure 4.6, as well as the supplementary video clip (PIC_LWFA(chap4).mp4). The PIC simulation was carried out with the OSRIS code, which was developed and is maintained by the Osiris Consortium, which consists of the Instituto Superior Técnico (IST) in Portugal and The University of California, Los Angeles (UCLA) in the United States. I'd like to thank Dr I Jong Kim and Dr Il Woo Choi who developed the bulk-glass and ultrathin-foil targets shown in Figure 3-13. I'd also like to thank Prof. Kyung Taec Kim and Mr. Yong Jin Kwon for providing the relativistic high-order harmonics animation (RHHG_CoReLS(Chap5).GIF).

# Author biography

## Hyung Taek Kim

**Hyung Taek Kim** is a head research scientist in the Advanced Photonics Research Institute (APRI), GIST. He got a Ph.D. from the Korea Advanced Institute of Science and Technology in 2004. Since joined Ultrashort Quantum Beam Facility, APRI in 2005, he has studied various applications of intense laser pulses, such as high-order harmonics, x-ray lasers, x-ray imaging, filamentation, laser wakefield acceleration, and gamma-ray generations. He led laser wakefield acceleration experiments as a research fellow at Center for Relativistic Laser Science (CoReLS), Institute for Basic Science (IBS). He has been the director of the Research Center for Plasma Applications with Ultraintense Lasers (PAUL center), APRI, since 2022.

**IOP** Publishing

Laser–Plasma Accelerators and Radiation Sources

**Hyung Taek Kim**

# Chapter 1

## Introduction

The term 'laser' stands for 'Light Amplification by Stimulated Emission of Radiation.' Based on the theoretical work of Charles H Townes and Arthur Leonard Schawlow, Theodore Maiman built the first laser in 1960 at Hughes Research Laboratories [1, 2]. Lasers are based on the stimulated emission inducing light amplification. The stimulated emission was firstly conceptualized by Albert Einstein in 1916 [3]. Because stimulated emission produces photons with identical properties to incident photons, emitted photons are coherently added to incident photons. When a meta-stable excited state has a larger population than a lower level, a phenomenon known as population inversion, light can be amplified with temporal and spatial coherence. Due to the spatial coherence, lasers can propagate a long distance by collimation and be focused to a cross-section of approximately wavelength square. Because of the laser's excellent temporal coherence, it can resolve complex atomic or molecular energy states near the ionization threshold with extremely narrow bandwidth. In contrast, using a broad-band laser medium and the superb temporal coherence, ultrashort laser pulses with pulse durations on the order of the laser wavelength can be realized.

Laser–plasma interactions have been studied since the 1960s, shortly after the invention of the laser. In the 1970s and 1980s, studies of laser-produced hot, dense plasmas for nuclear fusion dominated laser–plasma research. Large nanosecond high-energy laser facilities, such as the Nova laser [4] at Lawrence Livermore National Laboratory in California and the GEKKO laser [5] at Osaka University in Japan, were constructed in the late 1970s. Nobel laureates D Strickland and G Mourou published a paper on chirped pulse amplification (CPA) [6] in 1989 that paved the way for the development of compact, ultra-intense lasers. Initiated by the ultrashort high-intensity lasers, laser–plasma research expanded in the 1990s to include a broad range of applications.

Plasma refers to an ionized gas, while the term 'plasma' was derived from medical science to refer to the transparent liquid in blood after corpuscles have been removed. Johann Purkinje (1773–1828), a Czech medical scientist, coined the term πλασμα,

doi:10.1088/978-0-7503-2703-9ch1  

which means 'moldable substance' or jelly in Greek. The Nobel prize laureate Irving Langmuir first used this term to describe an ionized gas in 1927 [7]; Langmuir was reminded of the way blood plasma carries red and white corpuscles when considering how an electrified fluid carries electrons and ions. Plasma is a state of matter, often indicated as the fourth state of matter, that is characterized by the presence of a significant portion of charged particles in any combination of ions or electrons. Plasma is the most abundant form of ordinary matter in the universe. Plasma can be generated by ionizing neutral gases by heating them to extreme temperatures or subjecting them to a strong electromagnetic field.

Since the 1990s, significant progress in laser-produced plasma studies has been made through the development of coherent ultrashort extreme-ultraviolet sources, such as high-order harmonic generation and laser–plasma x-ray lasers. In the 2000s, laser intensity reached the relativistic regime, and scientists have continued to investigate new ways to use laser-produced plasmas, such as the development of laser wakefield accelerators, ion acceleration, betatron radiation, and inverse Compton scattering. Overall, laser–plasma research has played a key role in advancing our understanding of the fundamental physics of plasmas and their potential applications in numerous research fields, such as nuclear physics, particle physics, material science, biology, etc. In recent years, petawatt laser pulses have enabled new regimes of particle accelerations and radiation sources that can provide particle and radiation sources with high energy and ultrashort pulse duration for a variety of applications. Therefore, I believe now is the time to study the interactions between ultra-intense lasers and plasmas in order to develop smart and powerful sources.

This book covers the most recent developments in ultra-intense laser and secondary sources resulting from laser–plasma interactions with ultra-intense lasers. This book will not cover the fundamentals of optics, quantum optics on laser science, and plasma physics, as there are numerous books on these topics. This book aims to impart essential knowledge on designing, performing, and analyzing ultra-intense laser–plasma interactions. Consequently, this book is structured as follows:

- Chapter 1: Introduction
- Chapter 2: Ultrashort intense lasers
- Chapter 3: Instrumentations for laser–plasma interactions with ultra-intense lasers
- Chapter 4: Laser particle accelerators
- Chapter 5: Radiation sources from laser–plasma interactions
- Chapter 6: Applications of laser–plasma sources

Each chapter contains an introduction, sections, and subsections with detailed descriptions of each topic. In addition, simple chapter-related questions and a list of references will be included at the end of each chapter. Then, let us begin to illuminate our minds with lasers and plasmas.

## References

[1] Rigby P 2010 And then there was light—Physics World (https://physicsworld.com/a/and-then-there-was-light/)

[2] APS News 2003 This month in physics history (https://aps.org/publications/apsnews/200312/history.cfm)

[3] Einstein A 1916 Emission and adsorption of radiation in quantum theory *The Collected Papers of Albert Einstein* (Princeton: Princeton University Press) The Berlin Years: Writings, 1914–1917 6 212 (https://einsteinpapers.press.princeton.edu/vol6-trans/224)

[4] Simmons W n.d. The nova laser fusion facility defense programs.

[5] Kato Y, Mizumoto Y, Yoshida K, Ueda K, Monma M and Yamanaka C 1979 GEKKO IV glass laser system design parts *Operation Rev. Laser Eng.* **7** 35

[6] Strickland D and Mourou G 1985 Compression of amplified chirped optical pulses *Opt. Commun.* **56** 219

[7] Fitzpatrick R 2021 Brief history of plasma physics (https://farside.ph.utexas.edu/teaching/plasma/lectures1/node4.html)

**Hyung Taek Kim**

# Chapter 2

## Ultrashort intense lasers

The acronym 'laser' stands for 'light amplification by stimulated emission of radiation,' and it was invented by Theodore Maiman in 1960. Since the first laser, laser intensity has rapidly increased, and a revolutionary technology for ultra-intense laser was discovered in 1987 by D Strickland and G Mourou, which was the chirped pulse amplification (CPA) method. The technology of CPA has enabled researchers to investigate a new regime of laser–plasma interactions in relativistic optics. Knowledge of ultra-intense lasers is required to comprehend the interaction between ultra-intense lasers and plasmas. The important technologies for composing ultra-intense lasers and characterizing ultra-intense lasers are described in this chapter.

Ultrashort intense lasers are the indispensable research tools for laser–plasma physics. Since the invention of the laser in the 1960s, the laser's power has increased continuously. The laser power can be defined by two terms: average power and peak power. Average power refers to the energy in average per unit time and is commonly used for continuous lasers or lasers with a high repetition rate. Peak power for pulsed lasers refers to the maximum instantaneous power in a laser pulse, which is conventionally defined as the energy of a pulse divided by the pulse duration. As depicted in figure 2.1, pulsed laser technologies such as Q-switching [1, 2], mode-locking [3], and CPA [4] substantially increase the peak power of lasers [5, 6].

As the peak power of lasers has increased, ultrashort intense laser technology has made possible new laser–matter interactions. When the laser intensity exceeds $10^{13}$ W cm$^{-2}$, the electric field strength of the laser pulse exceeds the Coulomb potential for the outermost electrons of atoms, ionizing the atoms in the laser field to form a plasma. When the laser intensity is increased beyond $10^{18}$ W cm$^{-2}$, the electrons' quiver motion in the laser field can approach the speed of light, and the laser–plasma interactions become relativistic [5, 7, 8]. To investigate laser–plasma interactions in different laser intensity regimes, it is necessary to comprehend ultrashort laser technology. In this chapter, therefore, the fundamental principles and techniques of

doi:10.1088/978-0-7503-2703-9ch2

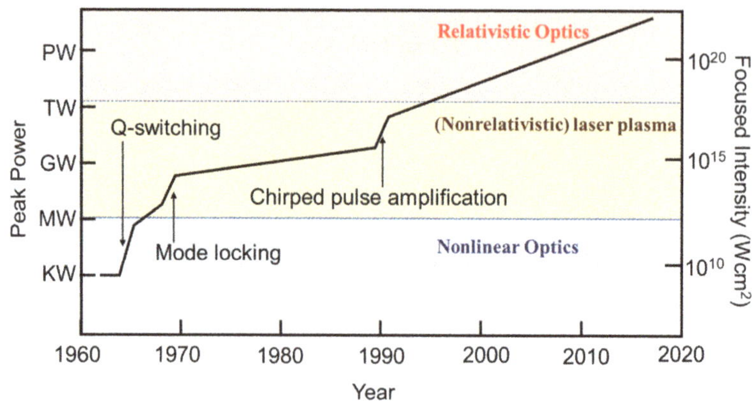


**Figure 2.1.** Advancement of ultrashort intense lasers.

ultrashort intense lasers for the study of laser–plasma-based particle accelerations and radiation sources are presented. This chapter is divided into three sections: short pulse generation and amplification methods, the temporal characterization and manipulation of ultrashort lasers, and laser focus control techniques.

## 2.1 Short pulse generation and amplification methods

In several laser facilities, ultrashort high-power lasers with peak powers exceeding one petawatt (PW, $10^{15}$ watt) have been constructed [9]. Future high-power lasers are anticipated to have a peak power of approximately 100 PW. How large is the PW scale? The total solar power on Earth is approximately 200 PW, and the heat flux between the equator and the poles is approximately 4 PW. How can the enormous peak power be achieved? Because the peak power is the pulse energy divided by the pulse duration. When a laser pulse has one J of energy and a duration of one femtosecond (fs, $10^{-15}$ s), its peak power is 1 PW. The femtosecond timescale is a duration of time comparable to the atomic and molecular motions. Therefore, a PW laser pulse signifies earth-scale power within the atomic time scale. To produce laser pulses with such a high peak power, it is necessary to generate ultrashort laser pulses. This section describes well-known techniques for generating and amplifying ultrashort laser pulses.

### 2.1.1 Q-switching

Q-switching is the first developed short-pulse high-power laser technology, which was demonstrated in early 1960s [1]. Briefly speaking, Q-switching is a tool to extract stored energy in a laser cavity for a short time by switching the Q-factor of the cavity. The Q-factor is defined by $2\pi$ times the ratio of the stored energy in the cavity to the energy dissipation per oscillation cycle, which is proportional to the ratio of the stored energy in the cavity and the cavity loss. The Q-factor can be varied by manipulating cavity losses using a light modulator incorporated into the optical cavity. The Q-switching lasers consist of the optical cavity, lasing medium, pumping

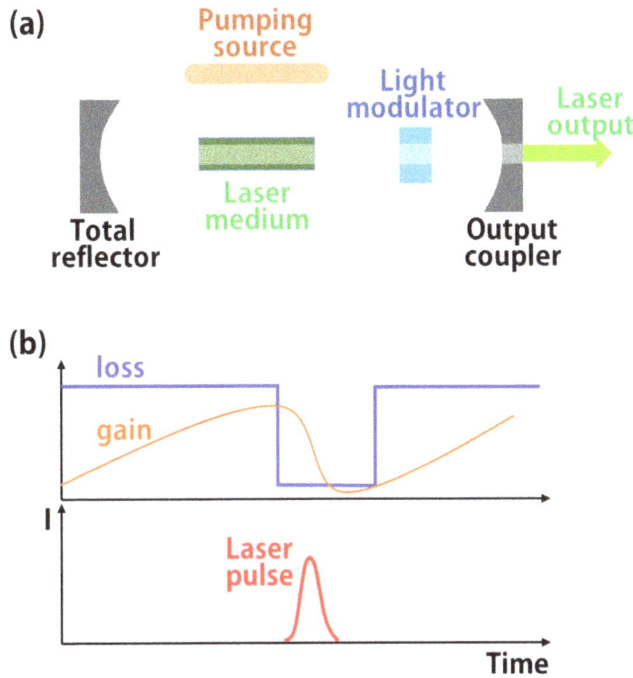


**Figure 2.2.** (a) Layout of Q-switching laser and (b) working principle of Q-switching laser.

source, and light modulator, as shown in figure 2.2(a). The light modulator can manipulate the Q-factor of the laser cavity by varying the loss of the cavity.

The Q-switching lasers form a high-intensity short-pulse laser by gating the loss of the laser cavity. Initially, the loss of the cavity is set to be high, i.e. the Q-factor is low, and the gain of the laser cavity increases in time. When the loss of the cavity is high, the stimulated emission is suppressed by disturbing the feedback of light in the cavity. The suppression of the stimulated emission by high cavity loss and the pumping power larger than the spontaneous emission lead to growth in the population inversion of lasing transition, which increases the gain of the laser medium. Then, the loss is reduced at a certain moment, i.e. the Q-factor of the cavity rises up, the stimulated emission is suddenly triggered, and strong lasing occurs up to the moment of the depletion of the population inversion. Due to high population inversion, the rapidly growing stimulated emission forms a laser pulse with high pulse energy and short pulse duration. Consequently, the switching of the cavity loss or cavity Q-factor produces a high-peak-power laser pulse.

To obtain the high-peak-power laser pulse with the Q-switching technique, the laser medium must have long upper-state lifetime for a high energy storage capability, high density of laser-active ions or atoms, and non-excessive gain efficiency to avoid strong amplified spontaneous emission. The rare-earth-doped crystals and glass, such as Nd:YAG, ND:$YVO_4$, Yb:YAG, Er:YAG, Nd:Glass, etc, are commonly used for the laser medium for the Q-switching laser. There are two types of light modulators for Q-switching; one is active Q-switching and the other is passive Q-switching. For the active Q-switching acousto-optic modulators are mostly used. During radio-frequency (RF) power application to the acouto-optic modulator, the modulator diffracts light and the loss of the cavity is high. When the power is off, the cavity loss drops down and strong lasing occurs suddenly. Saturable absorbers are typically used for the passive Q-switching. The saturable absorber is an optical component that has decreasing optical absorption at high optical intensities. In passive Q-switching, the amplified pulse is automatically formed when the gain has reached a sufficiently high level to induce the loss reduction of the saturable absorber. The Q-switching lasers mostly have pulse durations in nanosecond ($10^{-9}$ s) range, with the peak power several orders of magnitude higher than that of continuous lasers, and the energy per pulse ranging from mJ–kJ. The Q-switching is usually operated repetitively with repetition rate typically in the range of 1–100 kHz.

### 2.1.2 Mode-locking

Mode-locking is a technique that can provide extremely short laser pulses [10, 11] having pulse durations on the order of femtosecond ($10^{-15}$ s). Lasers have natural gain bandwidth and the laser pulse can have the pulse duration inversely proportional to the bandwidth when the longitudinal modes of the laser's cavity have a fixed-phase relationship. Mode-locking is the technique of inducing the fixed-phase relationship between the longitudinal modes of the laser's resonant cavity to form a train of ultrashort pulses. For a simple laser cavity composed of a laser medium and two planar mirrors separated by $L$, the frequency separation of adjacent longitudinal modes is given by $c/2L$, where $c$ is the speed of light. If there is one longitudinal mode in the bandwidth of the laser medium, pure single frequency laser or single-mode laser can be made. When there are several or lots of modes in the laser bandwidth, so-called multi-mode laser, longitudinal modes interfere with each other because they are all coherent and spatially overlapped.

The concept of mode-locking can be understood easily by the superposition of the longitudinal modes that are separated equally in frequency domain. Figure 2.3 shows an example of the superposition of seven longitudinal modes in a spectrum of Gaussian profile. When the longitudinal modes have a random-phase relationship, the electric field from the laser cavity is random noise signal. In contrast, the superposition of the longitudinal modes forms a train of ultrashort pulses when the longitudinal modes have a definite-phase relationship. The time interval between the pulses is given by the round-trip time of the pulse, $2L/c$. The minimum achievable pulse duration, called transform-limited pulse duration, is inversely proportional to the bandwidth of the laser spectrum, when the spectrum includes abundant longitudinal modes. Thus, the transform-limited pulse duration is given by

$$\tau_l = \frac{\alpha}{\Delta \upsilon} = \frac{2\pi\alpha}{\Delta\omega}, \tag{2.1}$$

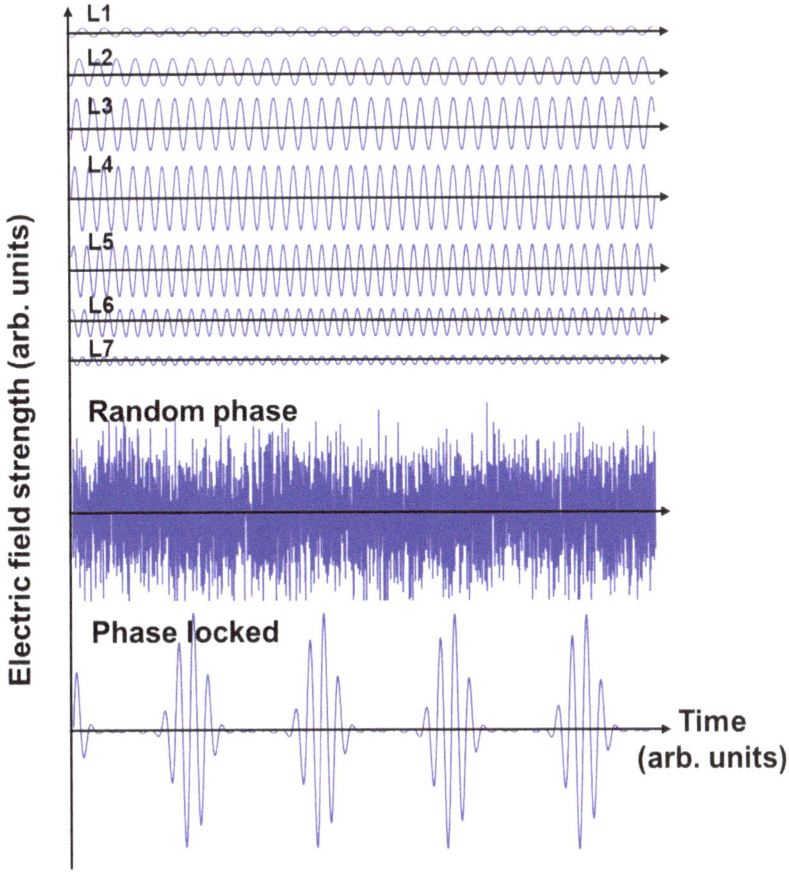


**Figure 2.3.** Concept of mode-locking. L1–L7 are each longitudinal mode from a laser cavity. The random-phase and phase-locked cases show the superposition of the longitudinal modes with random phases and a fixed-phase relationship, respectively.

where the $\Delta \upsilon$ is the bandwidth in frequency domain in units of Hz and $\Delta \omega$ is bandwidth in angular frequency domain in units of radian $s^{-1}$. The $\alpha$ is the time-bandwidth product that depends on the pulse shape. For instance, $\alpha$ is 0.44 for Gaussian pulse shape and 0.315 for a hyperbolic-secant-squared ($sech^2$)-shaped pulse. Therefore, broader bandwidth of laser medium supports shorter pulse duration for mode-locked laser.

The mode-locking also can be operated in an active or passive way. The structure of optical cavity is similar with Q-switching composed of optical cavity, laser medium, and optical modulator for active mode-locking or saturable absorber for passive mode-locking. For the active mode-locking, the light modulation should be the same as the round-trip time of the optical cavity. One of the well-known mode-locking methods is Kerr-lens mode-locking [12–15]. Kerr-lens mode-locking is a kind of passive mode-locking via Kerr nonlinear effect, i.e. Kerr-lens effect.

The optical Kerr effect [16–19], or AC Kerr effect, is an optical nonlinear effect by the refractive index variations proportional to the instantaneous light irradiance. As the laser intensity increases, the optical crystal or laser medium do not have any more constant refractive index due to the increasing contribution of the nonlinear polarization, i.e. the third-order susceptibility ($\chi^{(3)}$). Thus, the refractive index changes as follows:

$$n = n_0 + n_2 I, \tag{2.2}$$

where $n_2$ is Kerr coefficient or nonlinear refractive index in SI units that is given by

$$n_2 = \frac{3\chi^{(3)}}{4\varepsilon_0 c\, n_0\, \mathrm{Re}(n_0)}. \tag{2.3}$$

The Kerr effect is mostly negligible because Kerr coefficient ($n_2$) is much smaller than the $n_0$. For most glasses and optical crystals, Kerr coefficient is in the order of $10^{-16}$ cm$^2$ W$^{-1}$, the laser intensity over gigawatt cm$^{-2}$ can cause non-negligible effects. For instance, a laser pulse with Gaussian spatial profile induces a higher refractive index at the center that can act as a convex lens. This transverse refractive index variation due to the Kerr effect makes the laser beam be focused by itself, which is a well-known phenomenon called self-focusing.

This Kerr-lens effect can cause passive mode-locking by selecting higher-intensity part in the intensity variation in continuous wave (CW) operation. As shown in figure 2.4(a), the laser cavity for Kerr-lens mode-locking is composed of laser medium, mirror pair, and aperture. A Gaussian-like laser-beam profile is a preferable condition for the Kerr-lens mode-locking. As lasing starts, the central

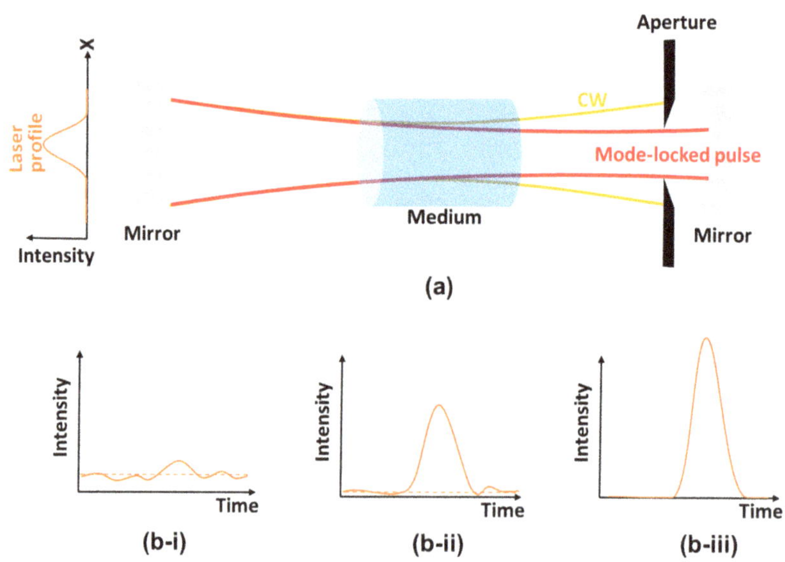


**Figure 2.4.** (a) Concept of Kerr-lens mode-locking in a laser cavity. (b) Temporal evolution of laser pulse profile during Kerr-lens mode-locking process. (i), (ii), (iii) are indices for time evolution.

laser intensity at the medium increases and goes near the threshold for the Kerr effect. When the laser intensity in time fluctuates and some part of laser intensity is over the threshold for the Kerr effect, the portion of the laser beam induces self-focusing and passes through the aperture without loss, while the CW part suffers the attenuation by the aperture. As the round trip of the laser in the cavity goes on, the self-focused part intensifies and forms a well-defined pulse with the time separation corresponding to the round-trip time of the laser cavity. The formation of well-defined pulse indicates the constructive interference between longitudinal modes. The pulse duration of the Kerr lens mode-locking is mainly determined by the bandwidth and the dispersion of the gain medium. A prism pair is often used to compensate the dispersion to obtain shortest pulse for a given laser medium and cavity structure.

### 2.1.3 CPA technique

CPA is a technique to amplify the ultrashort laser pulses to have very high peak power without optical damages in the amplification chains. When ultrashort laser pulses are amplifying, the peak power of the pulse increases rapidly in the laser amplifier and the optical components in the amplifier can be damaged easily. A solution to avoid the optical damages in the amplifiers is enlarging the beam size in space to reduce the peak intensity. However, the laser system gets bulky and costly as the laser beam size increases. An alternative and very efficient way, the CPA technique, was demonstrated by Donna Strickland and Gérard Mourou at the University of Rochester in 1985 [4]. They received the Nobel Prize in Physics in 2018 for the groundbreaking invention that enabled amplifying ultrashort laser pulses to be extremely high peak power over PW ($10^{15}$ watt) [9] without the problems of optical damages.

The CPA technique in principle uses extension of laser pulse duration in time, stretching the pulse to reduce the peak intensity in the amplification stages. Figure 2.5 shows the scheme of the CPA technique. Usually, femtosecond laser pulse is generated from the laser oscillator by the mode-locking process. The ultrashort low-energy laser pulse is then stretched in time by a pulse stretcher with diffractive optics. The pulse stretcher commonly consists of a grating pair that makes longer-wavelength part, i.e. lower-frequency part, travel shorter path than the shorter-wavelength part. Thus, the laser pulses after the pulse stretcher have positive laser chirp, i.e. positive group delay dispersion (GDD). The ultrashort laser pulses with the pulse duration of tens of femtoseconds are stretched to have pulse duration of sub-ns to several ns. The stretched pulse is amplified by the laser amplifiers to obtain high-energy stretched laser pulses. Then the amplified laser pulse is compressed by the reverse process of the pulse stretcher and the laser pulse duration returns to its original value. After the pulse compression, the peak power of the laser pulse becomes extremely high. The CPA technique allows construction of PW lasers and the peak intensity of focused laser pulse reaches $10^{23}$ W cm$^{-2}$ [6]. The supplementary video (PW_laser_UQBF.mp4, available to download from https://doi.org/10.1088/978-0-7503-2703-9) shows the operation scheme of a PW laser

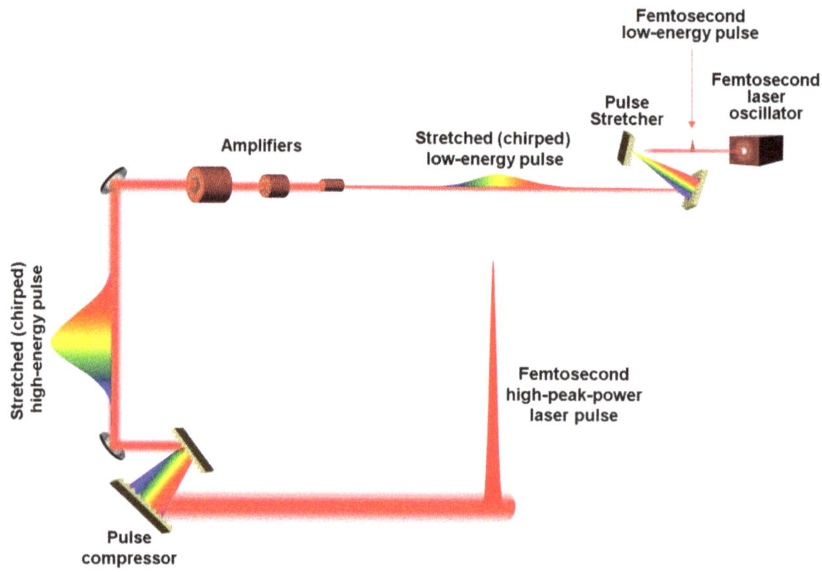


**Figure 2.5.** Schematic layout of CPA.

installed at Ultrashort Quantum Beam Facility (UQBF) at Advanced Photonics Research Institute, Gwangju Institute of Science and Technology [20–22]. In the pulse compressor, the laser intensity can induce nonlinear effects that can deform the spatio-temporal structure of the laser pulse. Thus, the pulse compressor and following optical systems are mostly installed in vacuum chambers. The CPA technology triggered the rapid increase of laser peak power and focused laser intensity since the 1990s.

## 2.2 Characterization and manipulation of ultrashort lasers

The intensity of a laser pulse is a key parameter to determine the laser–plasma interactions. To characterize the intensity of a laser pulse, the pulse energy, pulse duration, and focused spot size of the laser pulse should be characterized. Several methods, such as photo diode, pyroelectric sensor, thermal detector, and so on, can be used to measure the energy of a pulse or average power. A pulse's energy is equal to the average power divided by the repetition rate. The average power meter should be chosen with the damage threshold and beam diameter in mind. Since electronic measurements are limited to ns or sub-ns, measuring the duration of ultrashort pulses is a challenging task. Understanding temporal pulse characterization is essential for calculating the pulse peak power and laser intensity. Furthermore, the spatial phase of the laser pulse is a crucial factor in laser focusing. This section discusses measuring and manipulating the spatial and temporal properties of laser pulses.

### 2.2.1 Temporal pulse characterization methods

The pulse duration of the laser pulse below the speed of electronic circuit cannot be measured by photodiode or any other electronic detectors. Thus, the temporal characteristics of ultrashort laser pulses should be measured by optical methods by correlating ultrashort pulses with nonlinear conversions. The electric field of ultrashort laser pulses in time can be expressed by

$$E(t) = A(t) \times e^{-i[\omega_0 t - \varnothing(t)]} \tag{2.4}$$

where $A(t)$, $\omega_0$, and $\varnothing(t)$ are, respectively, the carrier frequency, amplitude, and phase of the laser pulse in time. The laser intensity profile in time can calculated by $I(t) = E(t) \times E^*(t)$. The electric field of the laser pulse in spectral domain is defined as the Fourier transform of the laser field in time as below:

$$E(\omega) = \int_{-\infty}^{\infty} E(t) \times e^{-i\omega t} dt = \int_{-\infty}^{\infty} A(t) \times e^{-i(\omega-\omega_0)t-\varnothing(t)}\, dt \equiv A(\omega) \times e^{-i\varphi\,(\omega-\omega_0)} \tag{2.5}$$

where $A(\omega)$ and $\varphi(\omega)$ are the spectral amplitude and spectral phase. The spectrum of a laser pulse is given by $I(\omega)= E(\omega) \times E^*(\omega) = A(\omega)^2$. The spectral phase can be expressed by Taylor expansion around carrier frequency as below:

$$\varphi(\omega - \omega_0) = \varphi\Big|_{\omega_0} + \frac{\partial\varphi}{\partial\omega}\Big|_{\omega_0}(\omega - \omega_0) + \frac{1}{2}\frac{\partial^2\varphi}{\partial\omega^2}\Big|_{\omega_0}(\omega - \omega_0)^2 + \ldots + \frac{1}{n!}\frac{\partial^n\varphi}{\partial\omega^n}\Big|_{\omega_0}(\omega - \omega_0)^n. \tag{2.6}$$

The first term is absolute phase or carrier envelope phase, and the second term corresponds to instantaneous frequency. The third term is group delay, which denotes a linear change in frequency over time, and the subsequent terms are associated high-order dispersions such as third-order dispersion and fourth-order dispersion, etc. Since the temporal electric field of a laser pulse is the Fourier transform of the spectral electric field, temporal characterization can be accomplished in both the time and spectral domains. As it is extremely difficult to measure the temporal phase of a laser pulse, temporal characterizations of laser pulses typically employ temporal envelope measurements or spectral field reconstruction techniques.

**(1) Autocorrelator**

The simplest way to measure the duration of an ultrashort laser pulse is correlating the pulse itself, so-called autocorrelator. Autocorrelators for pulse characterization use second-harmonic (SH) process because the field correlation without nonlinear process only provides the Fourier transform of the laser spectrum described by Wiener–Khinchin theorem. By applying SH conversion, the autocorrelation of ultrashort laser pulse can provide information on the pulse duration. There are two kinds of autocorrelators; one is intensity autocorrelator and the other is interferometric autocorrelator [23–26].

Figure 2.6(a) shows the typical optical setup of the intensity autocorrelator. A laser pulse is divided into two replica of the pulse and the time delay of the two pulses is controlled by detuning the path length of one pulse. Two pulses are

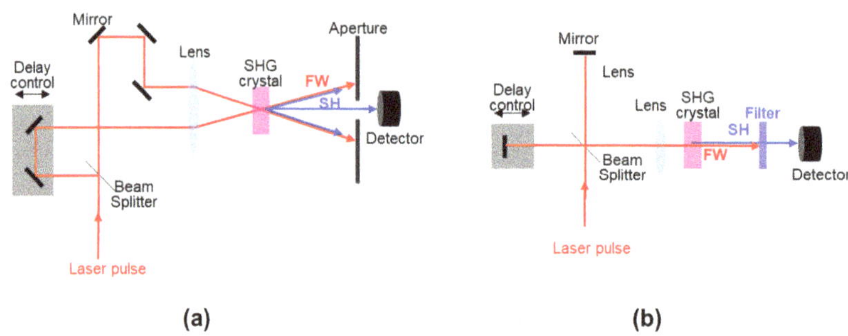


**Figure 2.6.** Optical layout of (a) intensity autocorrelator and (b) interferometric autocorrelator. FW indicates fundamental wave and SH is second-harmonic wave.

non-collinearly focused on an SH crystal by a lens. The mixed signals of fundamental and SH for each laser direction are blocked by aperture and the pure SH signal in the middle of each direction is detected by photodetector. The slow response time is not a problem because autocorrelation signal is a time-integrated signal for each time delay. The on-axial component of the second-harmonic generation (SHG) by two replicas provides only the intensity product. Thus, the intensity autocorrelation signal is described by the following formula:

$$I_{\mathrm{AC}}(\tau) = \int_{-\infty}^{\infty} I_L(t) \times I_L(t-\tau)dt, \tag{2.7}$$

where $\tau$ is the time delay between two replicated laser pulses, $I_{\mathrm{AC}}$ is autocorrelation, and $I_L$ is laser intensity profile. Figure 2.7(a) shows an example of an intensity autocorrelator signal that can be taken by the time-averaged SH signal with respect to time delay ($\tau$). The intensity autocorrelator does not provide the laser pulse profile, but it does provide the pulse duration with an assumption about the shape of the laser pulse profile. The pulse duration for sech$^2$-shaped pulses is 0.65 times the width of the autocorrelation signal, while the conversion factor depends on the pulse shape.

Intensity autocorrelation does not require a fast-response photodetector but measuring ultrashort laser pulses with durations below 20 fs presents difficulties. In the measurement of ultrashort pulses, chromatic aberrations, bandwidth of optical components, and group delay mismatch in SH crystal, geometric smearing effects caused by the incidence angle of two replicas to the SH crystal can lengthen the intensity autocorrelation signal, resulting in a measured pulse duration longer than the actual pulse duration. Since the signal disappears for large time delays, the intensity autocorrelation is referred to as background-free, whereas this is not the case for interferometric autocorrelators, as will be discussed below.

The interferometric autocorrelator uses the SH of the superposed field of the two replicated laser pulses, whereas the intensity autocorrelator is based on the product of intensity. Since the interferometric autocorrelator employs the superposition of two fields, the two replicas must be incident collinearly to the SH crystal. Due to collinear incidence, the geometric smearing effect and mismatch

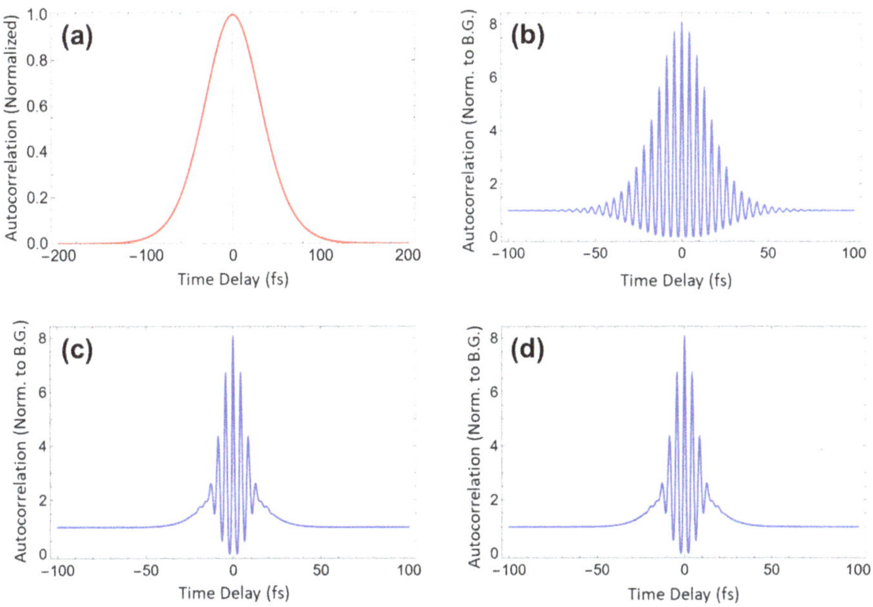


**Figure 2.7.** Autocorrelation signals. (a) Intensity autocorrelation signal for a laser pulse of sech$^2$ intensity profile with pulse duration in full width at half maximum (FWHM) of 50 fs. (b) Interferometric autocorrelation signal for a laser pulse of sech$^2$ intensity profile with 20 fs duration in FWHM. Panels (c) and (d) are interferometric autocorrelation signals that have the same profile as (b), but positive and negative laser chirp, respectively. For all cases, the wavelength of the laser field is 800 nm.

problem are not problematic in interferometric autocorrelators, which enable shorter pulse measurements than intensity autocorrelators. The optical layout of interferometric autocorrelator, based on Michelson interferometer, is shown in figure 2.6(b). The interferometric autocorrelation signal can be explained by following equation:

$$I_{AC}(\tau) = \int_{-\infty}^{\infty} |\,(E_L(t) + E_L(t - \tau))^2|^2 \, dt, \tag{2.8}$$

where $E_L$ is a complex electric field of laser pulse. $E_L(t)$ can be defined by $A(t)e^{i\omega t}$, where $A(t)$ is amplitude and $\omega$ is angular frequency of the laser field. As depicted in figure 2.7(b), the interferometric autocorrelation trace shows a rapid oscillation. Expansion of equation (2.8) can describe the profile of the interferometric autocorrelation

$$\begin{aligned} I_{AC}(\tau) = &\int_{-\infty}^{\infty} I^2(t) + I^2(t - \tau)dt + 4\int_{-\infty}^{\infty} I(t) \times I(t - \tau)dt \\ &+ 2\int_{-\infty}^{\infty} [I(t) + I(t - \tau)] \times E(t) \times E^*(t - \tau)dt + \int_{-\infty}^{\infty} E^2(t) \times E2^*(t - \tau)dt, \end{aligned} \tag{2.9}$$

where $I(t) = E(t)E^*(t)$. The initial term is constant background, and the second term is intensity autocorrelation. The third term is the interferogram of $E(t)$

oscillating at $\omega$, while the fourth term is the interferogram of the SH field oscillating at $2\omega$. According to equation (2.9), the minimum of the autocorrelation signal at a large time delay corresponds to two times the SH signal of the single pulse, whereas the peak of the autocorrelation signal corresponds to 16 times the SH signal. As a result, the ratio of maximum signal to background signal is 8:1, as shown in figure 2.7(b), indicating that the interferometric autocorrelator is properly aligned. The intensity autocorrelation can be extracted from the interferometric autocorrelation signal by low-pass filtering to remove oscillating terms (the third and fourth terms in equation (2.9)) and subtracting the constant background signal. With extracted intensity autocorrelation signal and a pulse profile assumption, it is possible to estimate the duration of a pulse.

The interferometric autocorrelation depends on the chirp of laser pulse. Thus, the pulse duration estimated by interferometric autocorrelator is not proper for chirped laser pulse, while the interferometric autocorrelator contains information on laser chirp. The interferometric autocorrelation signal of chirped pulse can be treated by improved post-processing method such as modified-spectrum auto-interferometric correlation [25]. However, the advanced pulse profile reconstruction methods based on interferometric autocorrelation have intrinsic limitation due to the symmetric signal of autocorrelation. The advanced interferometric autocorrelations method cannot distinguish between laser pulses with the same intensity profile and the same laser chirp with the opposite sign, as shown in figures 2.7(c) and (d), because the autocorrelation signals are identical. Moreover, the same interferometric autocorrelator can be produced with either an asymmetrically skewed pulse profile with zero chirp or a symmetric profile with complex chirps. In order to characterize ultrashort laser pulses, a precise method of measuring laser profile and chirp is necessary, which will be described below.

**(2) Frequency-resolved optical gating**

Since the autocorrelators cannot provide the precise measurement on the electric field of ultrashort laser pulses due to their intrinsic limitation of symmetric signal profile, proper methods to measure the electric field configuration, including spectral amplitude and spectral phases, are required. In order to solve this problem, Rick Trebino and Daniel J Kane invented frequency-resolved optical gating (FROG) [26–28]. The FROG method can measure the profile and the spectral phase of ultrashort laser pulses in broad time ranges from few femtoseconds to nanosecond. FROG method is based on the concept of spectrogram. In general, the spectrogram of an electric field can be written as below:

$$I_{\mathrm{SPECTROGRAM}}(\omega, \tau) = \left| \int_{-\infty}^{\infty} E(t) \bullet g(t-\tau) e^{-\mathrm{i}\omega t} dt \right|^2, \qquad (2.10)$$

where $g(t - \tau)$ is a variable-delay gate function and $\tau$ is the time delay between electric field and the gate function. The spectrogram function is often used to analyze the time-frequency analysis on an ultrashort broad-bandwidth laser field. The FROG method utilizes the spectrally resolved SH signal of autocorrelation as

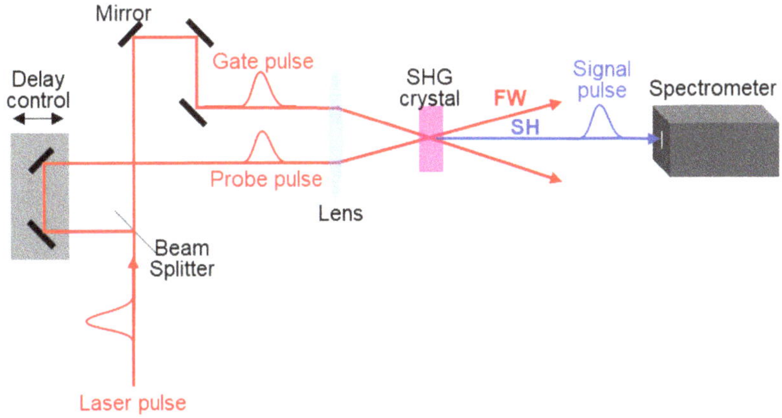


**Figure 2.8.** Optical layout of FROG method to characterize ultrashort pulses.

shown in figure 2.8 to form a kind of spectrogram signal by optical measurement. Thus, FROG in simplest version measures the spectrum at each time delay of two replicas for signal pulse and gate function both with the replicas. These spectral measurements of SH autocorrelation signal, often called SHG FROG spectrogram or SHG FROG trace, are given by the following equation:

$$I_{\text{SHG FROG}}\,(\omega,\,\tau) = \left| \int_{-\infty}^{\infty} E(t)\bullet E(t-\tau)e^{-\mathrm{i}\omega \mathrm{t}}dt \right|. \tag{2.11}$$

The output signal of FROG measurement is given by two-dimensional (2D) images, as shown in figure 2.9. Retrieval of the pulse from its FROG trace is accomplished by using a 2D phase-retrieval algorithm. That means iteratively tracing the original signal in time and frequency domains with proper physical constraints. 2D phase reconstruction was proven to be a unique solution. By this reconstruction algorithms, the intensity profile, and phase variations can be recorded by FROG method, even for very complex or extremely short pulses. As a results, nowadays, FROG is a well-established and the mostly used method for measuring ultrashort laser pulses.

**(3) Spectral phase interferometry for direct electric-field reconstruction**

When two laser pulses with a time delay enter a spectrometer, the spectrum of the two pulses is modulated, a phenomenon known as spectral interferometry (SI) [29], with an average spectral modulation frequency spacing inversely proportional to the time delay. The SI signal is given by the following formula:

$$I_{SI}(\omega) = I_{L1}(\omega) + I_{L2}(\omega) + 2\sqrt{I_{L1}(\omega)\times I_{L2}(\omega)} \times \cos(\varphi_{L1}(\omega) - \varphi_{L2}(\omega) - \omega\tau), \tag{2.12}$$

where $I_{L1}$ and $I_{L2}$ are spectral intensity of the two pulses, and $\varphi_{L1}$ and $\varphi_{L1}$ are the spectral phases of them.

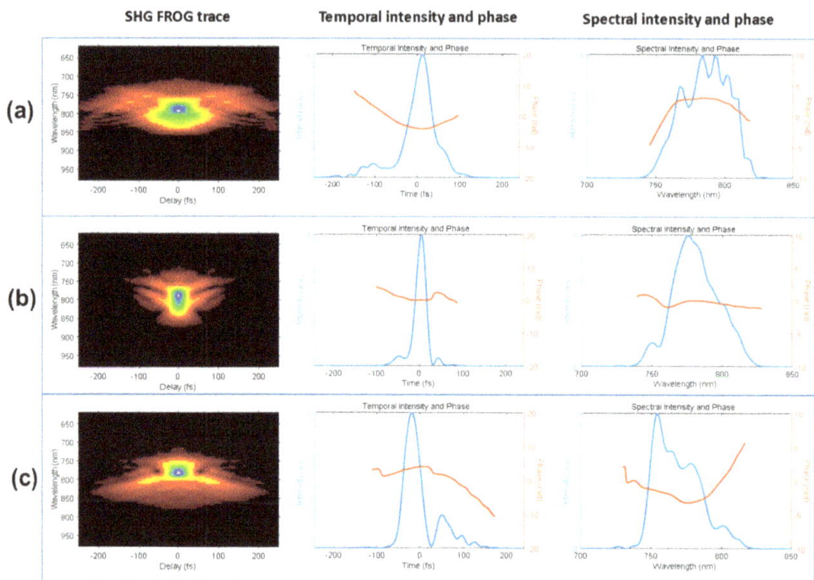


**Figure 2.9.** FROG trace and reconstructed temporal and spectral profile of a 27 fs laser pulse with (a) negative chirp (GDD = $-910$ fs$^2$, pulse duration= 54 fs), (b) chirp-free, and (c) with positive chirp (GDD = $+1190$ fs$^2$, pulse duration = 43 fs). The pulse durations are in full width at half maximum (FWHM).

The oscillating term of SI is dependent on the time delay ($\tau$) and spectral phase difference between two pulses ($\varphi_{L1}(\omega) - \varphi_{L2}(\omega)$). Thus, the SI can be used to precisely measure the time delay between two pulses and to characterize the spectral phases of an unknown laser pulse by comparing it with a known laser pulse.

Spectral phase interferometry for direct electric-field reconstruction (SPIDER) [30, 31] is an SI-based ultrashort pulse measurement technique. The SI signal of two time-delayed spectrally sheared laser pulses is used by SPIDER. The spectral shear in SPIDER measurements is produced by sum-frequency generation (SFG) of two replicas of an unknown pulse and a temporally stretched (chirped) pulse from a copy of the pulse itself. Because the replicas are converted to SFG at different stretched pulse frequencies, the two SFG pulses have the same intensity and phase with a central frequency shift, i.e. spectral shear, as shown in figure 2.10. The SI between the two SFG pulses allows the spectral phase at one frequency to be compared to the spectral phase at another frequency separated by spectral shear. The SI signal of two time-delayed pulses with spectral shear is written as follows:

$$I_{\text{SPIDER}} = I(\omega) + I(\omega - \Omega) + 2\sqrt{I(\omega) \times I(\omega - \Omega)}\cos(\varphi(\omega) - \varphi(\omega - \Omega) - \omega\tau). \quad (2.13)$$

As depicted in figure 2.10(b), the unknown spectral phase of the pulse can be extracted using simple calculations. First, the interferogram is Fourier transformed into the pseudo-time domain, and then one of the sideband peaks (alternating current (AC) term) is extracted. The argument of the inverse Fourier transform of an AC term yields the SPIDER phase, $\theta(\omega) = \varphi(\omega) - \varphi(\omega - \Omega) - \omega\tau$. The reference term ($\omega\tau$) can be subtracted from the SI measurement by measuring the time delay between the

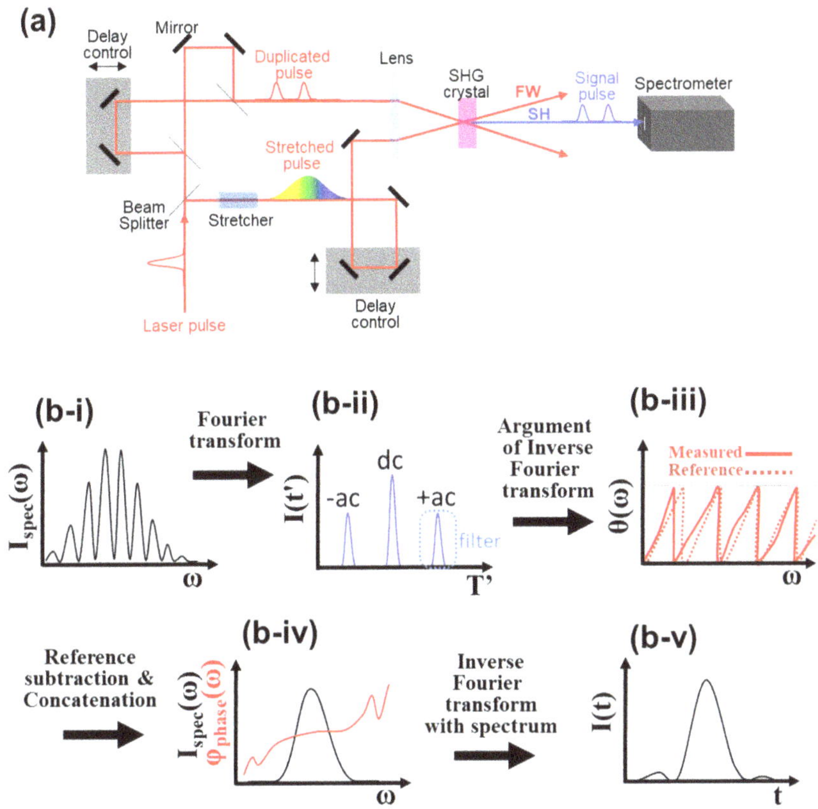


**Figure 2.10.** (a) Optical layout of SPIDER. (b) Reconstruction process for temporal pulse profile from SPIDER signal.

duplicated pulses without spectral shearing. Then, the spectral phase of the pulse can be determined by concatenating the SPIDER phase with the following formula:

$$\varphi\,(\omega_0 + \mathrm{N}\,|\,\Omega\,|) = \begin{cases} -\sum_{n=1}^{N}\theta(\omega_0 + \mathrm{N}\Omega),\ \ \Omega > 0 \\ \sum_{n=0}^{N-1}\theta(\omega_0 + \mathrm{N}\,|\,\Omega\,|),\ \Omega < 0 \end{cases}, \tag{2.14}$$

given the integer N and the concatenation grid $\{\omega_0 + N\,|\,\Omega\,|\}$. The spectral intensity can be extracted from the direct current terms of SI signal without shearing or independently measured spectrum. The temporal profile and phase can be reconstructed utilizing inverse Fourier transform with the field amplitude and phase in spectral domain. SPIDER can provide faster reconstruction than FROG because it only deals with one-dimensional data, whereas FROG reconstructs a 2D image. In contrast, SPIDER may be more sensitive to noise and misalignment because SPIDER uses direct conversion from the measured data, whereas FROG can evaluate the conversion process to determine the quality of the reconstructed data.

### 2.2.2 Waveform control of ultrashort laser pulses

The waveform of a laser pulse indicates the shape of field oscillation in time, including amplitude, phase, and polarization. If the polarization of a laser field is fixed, the waveform of a laser pulse designates the temporal laser field as depicted in equation (2.4). Since the temporal electric field of a laser pulse is Fourier transform of spectral field, the waveform of a laser pulse for a given polarization can be equivalently defined in spectral domain. Therefore, the simplest way to manipulate the waveform of a laser pulse can be to modulate the spectrum by spectral filters or inducing laser chirp by dispersive optics. Laser chirp also can be controlled by tuning the grating distance in compressor for CPA lasers. However, these methods have limitations in making designed waveforms of laser pulses. Acousto-optic programmable dispersive filters (AOPDFs) [32, 33] can be employed to form a designed waveform of CPA ultra-intense lasers, while spatial light modulators can be used for low-power ultrashort laser pulses.

The AOPDF is a collinear-type acousto-optic modulator that modifies the phase and amplitude of ultrashort laser pulses. AOPDF makes use of the birefringent acousto-optic effect to control the phase and amplitude. An RF-excited transducer launches an acoustic wave into an acousto-optic birefringent crystal, creating an acoustic grating in the crystal. The incident laser pulse is initially polarized along the fast (ordinary) axis of the birefringent crystal. When a laser frequency is phase-matched to an acoustic grating after a certain propagation length, a portion of its energy is diffracted onto the slow (extraordinary) axis polarization. Thus, when the laser pulse exits the AOPDF, it is recomposed with all the spectral components on the extraordinary axis that were diffracted from the ordinary axis at various positions. Due to the different velocities of the two polarizations, each optical frequency will experience a different time delay. The acoustic power at the diffraction positions can control the amplitude of the output pulse or diffraction efficiency. Since the acoustic wave generation can be tuned by manipulating the RF signal, the AOPDF can generate a laser pulse with the desired amplitude and phase.

The AOPDF can be installed after the stretcher of a CPA laser system, as shown in figure 2.11, because the laser pulse's peak power is not yet very high. After

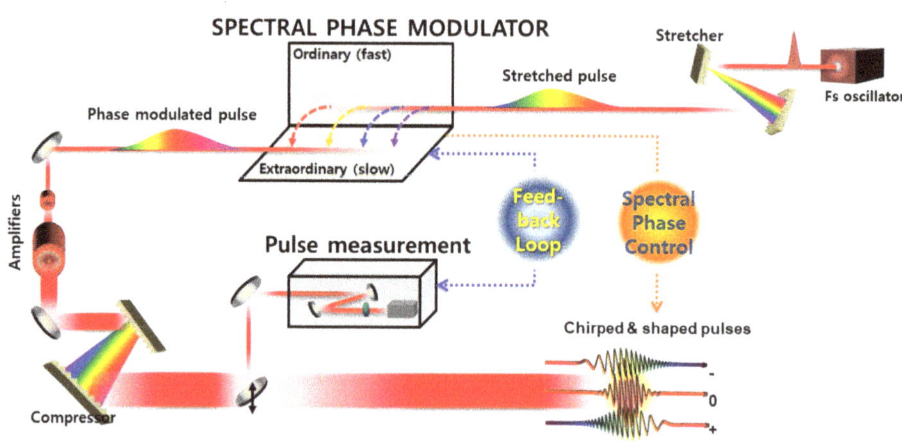


Figure 2.11. Scheme of waveform control technique for CPA laser pulses; reproduced from [34]. CC BY 4.0.

amplifiers and pulse compressors, the laser pulse waveform can be precisely characterized by FROG, SPIDER, or variations of the two methods. From the measured data, feedback controls to AOPDF can be applied to generate a laser pulse with the desired waveform. In addition, we can modify each of the spectral phase terms in equation (2.6) using AOPDF to examine the effects of each spectral phase term on laser–plasma interactions. Therefore, the waveform control method utilizing AOPDF is an indispensable technique for controlling laser–plasma interactions.

## 2.3 Laser focus control techniques

The beam waist size of a focused intense laser pulse is an important parameter to determine the peak intensity. Sections 2.1 and 2.2 describe generation, amplification, temporal characterization, and waveform control of intense laser pulses, which are mostly related to the temporal configurations of the laser pulses. This section will focus on the spatial properties of laser pulses that define the spatial qualities of the laser pulses.

### 2.3.1 Gaussian beam propagation

A Gaussian beam is a beam that has transversed amplitude profile of a Gaussian function. A laser beam from a simple resonator, composed of parabolic mirrors and homogeneous medium, has the lowest-order spatial mode ($TEM_{00}$) of Gaussian profile in paraxial approximation regime [35]. Thus, Gaussian beam propagation is basic to understanding laser beam propagation and focusing. The higher spatial modes of a laser beam from the simple resonator in paraxial regime can be described by Hermite–Gaussian modes. For a monochromatic plane wave, the Gaussian beam with propagation direction of $z$ and polarization axis in $x$ is given by:

$$E(r, z) = E_0 \frac{w_0}{w(z)} e^{\left(\frac{-r^2}{w(z)^2}\right)} e^{\left(-i\left(kz + k\frac{r^2}{2R(z)} - \psi(z)\right)\right)} \hat{x} \tag{2.15}$$

where $r$ is the radial distance from the center axis of the beam, $z$ is the axial distance from the beam's waist (or focus), $k$ is wavenumber, and $E_0$ is the electric field amplitude at the origin (or at the center of the beam waist). $w$ indicates the beam radius, radial distance from the beam center to the point where the amplitude decreases to $1/e$. $w_0$ and $w(z)$ are the beam radius at the beam waist and $z$ longitudinal position of $z$, respectively. $R(z)$ is the radius of wavefront curvature at $z$ and $\psi(z)$ is Gouy phase. The corresponding laser intensity distribution is given by

$$I(r, z) = E(r, z)E * (r, z) = I_0 \left(\frac{w_0}{w(z)}\right)^2 e^{\left(\frac{-2r^2}{w(z)^2}\right)}, \tag{2.16}$$

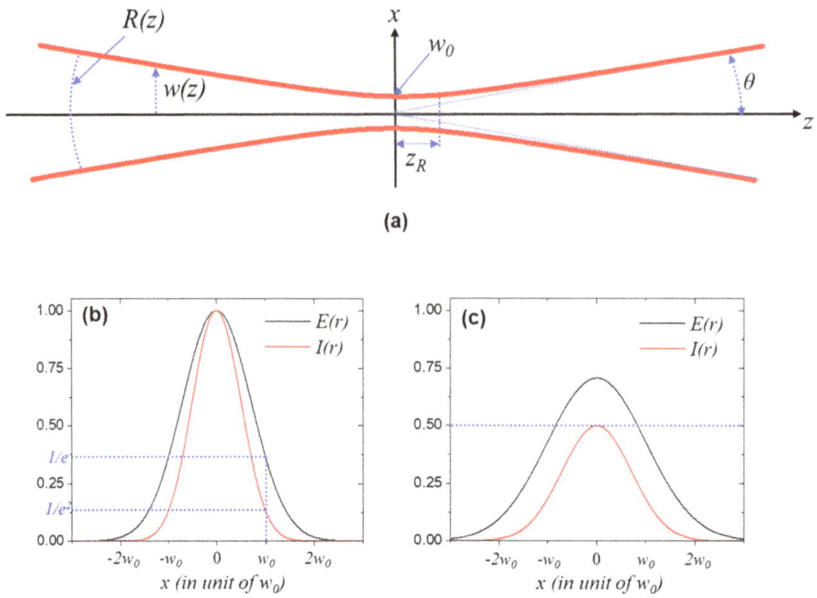


**Figure 2.12.** (a) Schematic drawing of Gaussian beam propagation. Gaussian beam profile (B) at the beam waist and (c) at $z$ position of Rayleigh length. Black and red curves are for electric field profile and intensity profile, respectively.

where the $I_0$ is the intensity at the center of the beam waist. Since equation (2.15) is the solution of wave equation for the paraxial approximation, the equation is valid when the beam waist size should be sufficiently larger than the wavelength ($w_0 \gg \lambda$).

As the laser beam is propagating, the beam size varies in the following equations:

$$w(z) = w_0\sqrt{1 + \left(\frac{z}{z_R}\right)^2}, \quad \text{where} \quad z_R = \frac{\pi w_0^2 n}{\lambda} \tag{2.17}$$

where $n$ is refractive index of medium where the Gaussian beam is propagating. $z_R$ is the Rayleigh length. After the propagating through Rayleigh length from the beam waist, the laser intensity of the Gaussian beam becomes 1/2 as shown in figure 2.12(c). The Rayleigh length defines the length range of focusing zone where the laser beam is not diverging significantly and the intensity of the laser pulse is intense enough to induce strong interactions. Twice the Rayleigh length is often called the confocal parameter. The asymptotic behavior of beam size is proportional to the propagation distance and the angle of the asymptotic line is given by

$$\theta = \lim_{z \to \infty} \arctan\left(\frac{w(z)}{z}\right) = \frac{\lambda}{\pi n w_0}. \tag{2.18}$$

The full divergence angle of the Gaussian beam at far field is $2\theta = 2\lambda/(\pi n w_0)$.

The radius of wavefront curvature and Gouy phase also can be written in terms of Rayleigh length as follows:

$$R(z) = z\left[1 + \left(\frac{z_R}{z}\right)^2\right] \tag{2.19}$$

$$\psi(z) = \arctan\left(\frac{z}{z_R}\right). \tag{2.20}$$

The radius of wavefront curvature, indicating flat wavefront, is infinite at far field ($z = \pm\infty$) and beam waist ($z = 0$). The wavefront of Gaussian beam is spherical in shape except at far field and beam waist position. The Gouy phase [36] is an additional phase shift that occurs in the propagation of a Gaussian beam across the focal region. The Gouy phase makes a difference in the propagation of focused Gaussian beam and plane wave. The physical origin of Gouy phase can be interpreted by transverse momenta spread and shift of expectation value of the axial propagation constant induced by transverse spatial confinement and uncertainty principle. The Gouy phase shifts from $\pi/2$ to $-\pi/2$ as a Gaussian beam propagating from $z = -\infty$ to $z = \infty$.

For simplifying the Gaussian beam description, we can employ complex beam parameter q given by

$$q(z) = z + iz_R. \tag{2.21}$$

By simple algebra, the electric field of a Gaussian beam can be written as

$$E(r, z) = \frac{1}{q(z)} e^{\left(-ik\frac{r^2}{2\,q(z)}\right)}. \tag{2.22}$$

The $q$ parameter has relation to radius of wavefront curvature and beam radius as follows:

$$\frac{1}{q(z)} = \frac{1}{R(z)} - i\frac{\lambda}{n\,\pi w^2(z)}. \tag{2.23}$$

The beam properties at a given propagation distance can be calculated using $q$ parameter and ABCD matrix in ray optics. For a given input beam $q$ parameter of $q_{in}$ and propagation ABCD matrix, the output $q$ parameter is given by

$$q_{out} = \frac{Aq_{in} + B}{Cq_{in} + D}. \tag{2.24}$$

The calculation $q$ parameter with ABCD matrix is a very useful tool because it can describe the Gaussian beam propagation with simple matrix calculations.

Even though the Gaussian beam is the fundamental spatial mode of a laser resonator, laser beams are not exactly Gaussian beam. Therefore, a measure is required to evaluate the laser beam mode similarity to the Gaussian beam. The most well-known parameter to indicate the similarity to Gaussian beam is $M^2$ factor [37, 38]. The $M^2$ factor is defined as

$$M^2 = \frac{\pi}{\lambda} w_{0M}\, \theta_M \geqslant 1 \tag{2.25}$$

where $w_{0M}$ and $\theta_M$ are, respectively, the beam waist size and the half-divergence angle for $M^2 > 1$. Only a Gaussian beam has $M^2$ factor of 1, while all other beams have $M^2$ factor larger than 1. The beam radius with $M^2$ factor is given by

$$w_M(z) = w_{0M}\sqrt{1 + \left(\frac{z}{z_{RM}}\right)^2} = w_{0M}\sqrt{1 + \left(\frac{z}{\pi w_{0M}^2 \lambda\, / M^2}\right)^2} \tag{2.26}$$

where $Z_{RM}$ is the Rayleigh length for $M^2 > 1$. For a given beam waist size, the Gaussian beam has the smallest possible beam divergence and real beam has larger divergence than Gaussian beam, while the Gaussian beam has the smallest possible beam waist size for a given beam divergence. The $M^2$ factor is an important parameter to define the spatial quality and focusability of a laser beam.

### 2.3.2 Wavefront of laser pulses

Wavefronts are surfaces connecting points with the same phase of a monochromatic wave. The wavefront of a laser pulse is an important parameter because it determines the quality of laser focus [39]. The wavefront aberration is defined by the deviation of wavefront from the perfect spherical wavefront [36]. For a circular shape beam, the wavefront aberration can be described by the Zernike polynomials. The Zernike polynomials are a sequence of polynomials that are orthogonal on the unit disk. The Zernike polynomials are a complete set in two variables of radial distance $\rho$ and azimuthal angle $\varphi$. There are even and odd Zernike polynomials defined by

$$\begin{cases} Z_n^m(\rho, \varphi) = R_n^m(\rho)\cos(m\varphi), & \text{even} \\ Z_n^{-m}(\rho, \varphi) = R_n^m(\rho)\sin(m\varphi), & \text{odd} \end{cases} \tag{2.27}$$

where $m$, $n$, and $n$–$m$ are non-negative integers. The radial distance is in the range of $0 \leqslant \rho \leqslant 1$ and the Zernike polynomials have values in between $-1$ and 1. The even and odd Zernike polynomial is even and odd function over the azimuthal angle $\varphi$. The radial Zernike polynomials for an even number of $n$–$m$ are given by

$$R_n^m(\rho) = \sum_{k=0}^{\frac{n-m}{2}} \frac{(-1)^k (n-k)!}{k!\left(\frac{n+m}{2} - k\right)!\left(\frac{n-m}{2} - k\right)!}\rho^{n-2k} \tag{2.28}$$

while it is 0 for an odd number of $n$–$m$. Specially, the radial function at $\rho = 1$ is defined as 1.

Figure 2.13 shows the low-order Zernike polynomials. The lowest term (zeroth order) is piston which indicates the mean value of a wavefront or phase profile across the pupil of an optical system. The first-order terms are $x$-tilt and $y$-tilt which indicate the transversal propagation directions of the laser beam. These low-order terms ($n$ = 0 and 1) of Zernike polynomials still indicate non-aberrated laser beam.

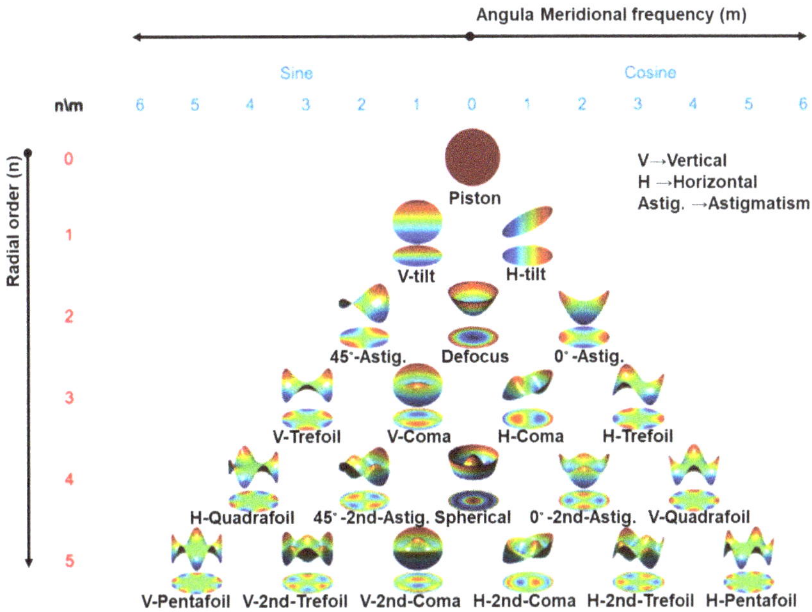


**Figure 2.13.** Surface plots of Zernike polynomials up to $n = 5$. Reproduced from [40]. © The Author(s). Published by IOP Publishing Ltd. CC BY 4.0.

The defocus ($n = 2$, $m = 0$) is the lowest order of optical aberrations in which the focus is not at the focal plane, or an image is simply out of focus. For the defocused beam, the wavefront is still perfect spherical shape, and the focus is at somewhere else. The other aberration term in the second order is astigmatism (figure 2.14 (b)). Astigmatism is the difference of laser focus in between two perpendicular transverse axes, which occurs usually when a laser beam is obliquely incident to the optics axis of focusing optics. The laser focus at one of the focal points of two axes is typically elliptic shape. With an astigmatism, a perfect focusing of a circular laser beam is no more possible. Coma (figure 2.14 (c)), in the third terms of Zernike polynomials, is an aberration in which a laser beam parallelly incident to optic axis, but off-centered, results in a trailing 'comet-like' blur.

Most noticeable aberration in 4th order is spherical aberration ($n = 4$, $m = 0$). The spherical aberration (figure 2.14 (d)) is an azimuthally symmetric aberration by the difference of focal length with respect to radial position $\rho$, which is caused when off-centered rays are refracted or reflected more or less than rays at the center. For the perfect focusing, focusing optics should be parabolic mirror or aspheric lens, while they are much more difficult to fabricate than spherical mirror or lens. When a single spherical concave mirror or convex lens is used, the spherical aberration is inevitable. As a result of spherical aberrations, the focus of beam elongates longitudinally significantly, as well as enlarging the focal radius. Figure 2.14 shows the spot diagrams for typical aberrations described above. In a real optical system, the aberration of the laser beam is more complex, while the aberrations of optical systems for monochromatic lights can be described by the combinations of Zernike polynomials.

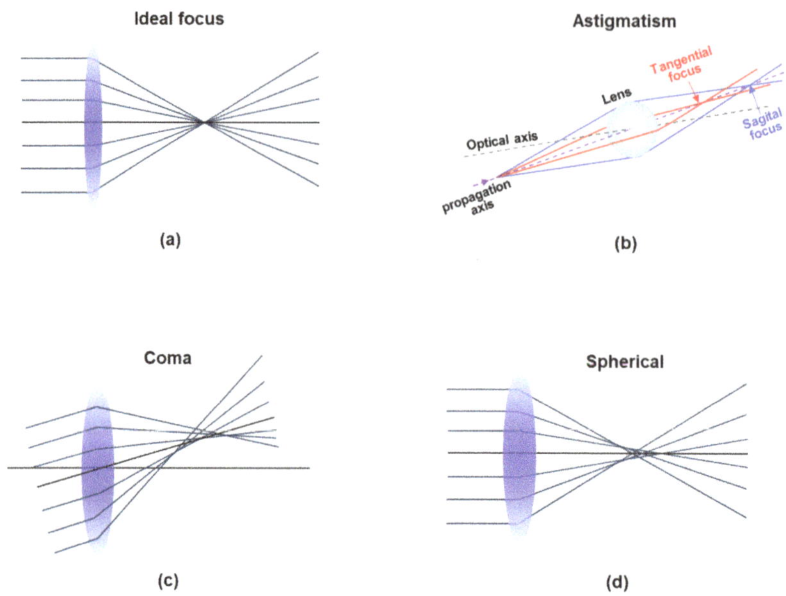


**Figure 2.14.** Conceptual illustrations of focusing of rays (a) without aberration, (b) with astigmatism, (c) with coma, and (d) with spherical aberration.

### 2.3.3 Measuring and controlling wavefront of laser pulses

The measurement and control of the wavefront of laser pulses are crucial effects on the focusing quality and the characteristics of laser–plasma interactions. Most common method to measure the wavefront is Shack–Hartmann sensor [42, 43], named after Johannes Franz Hartmann and Roland Shack. Shack–Hatmann sensor is a simple apparatus composed of a micro-lens array and a camera located at the focal plane of the micro-lens. For a plane wave without aberration, the Shack–Hartmann sensor detects an optical image of equally spaced dots (figure 2.15 (a)). When an aberrated beam is incident to Shack–Hartmann sensor, the dots are deviated from the original position, and the displacements of the dots correspond to the aberration averaged over the entrance area of the lenslet of the micro-lens. Thus, the wavefront of the incidence beam to Shack–Hartmann sensor can be reconstructed from the deviations of the positions of focused dots from the micro-lens. The other method is using diffractive grating [44] instead of micro-lens array, which can provide better spatial resolution than the Shark-Hartmann sensor.

The aberration of the laser beam can be minimized or manipulated by using deformable mirrors. The deformable mirrors [45–47], also known as adaptive optics, are mirrors capable of changing the shape of reflecting mirror surface (figure 2.15 (b)). The deformable mirror has been developed for large aperture space telescope on the ground to compensate the effects of atmosphere. As the beam sizes of the ultra-intense lasers are getting larger to achieve higher laser peak power, the compensation of the wavefront aberration has been becoming essential. There are several types of deformable mirrors such as bimorph type, mechanical actuator type, and membrane

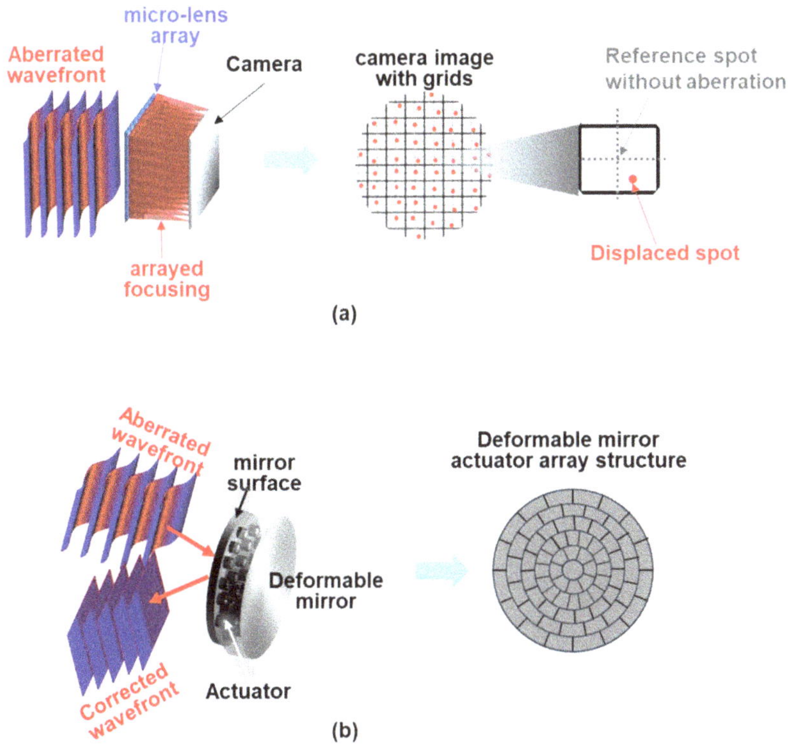


**Figure 2.15.** Schemes for (a) Shack–Hartmann wavefront sensor and (b) deformable mirror [41].

type and so on. Each of the deformable mirror types utilizes different method to make a controllable and programmable mirror surface. For optimizing the focusability of laser pulses, the wavefront aberration can be iteratively compensated by using wavefront sensor and deformable mirror. With adaptive optics system composed of a deformable mirror and a wavefront sensor, a laser pulse can be focused near the wavelength of the laser with short focal length off-axis parabolic (OAP) mirror with F#/1. In addition, the quality of the laser focus can be quantified by Strehl ratio defined by the ratio of maximum irradiance of the measured focal spot to the ideal maximum irradiance from a theoretical aberration-free focus. High focusability, or high Strehl ratio, of ultra-intense laser pulses is essential for laser–plasma interactions and adaptive optics systems can be an excellent solution for that.

### 2.3.4 Chromatic aberrations of ultrashort laser pulses

Since the wavefront sensors are mostly for a monochromatic light, chromatic aberrations cannot be compensated by the adaptive optics systems. For ultrashort high-power lasers based on CPA technology, the chromatic aberrations can occur by refractive optics in the beam lines or misalignments of diffractive optics in pulse stretcher and pulse compressors. The chromatics aberrations of transmissive optics or misalignments of diffractive optics can cause angular chirp, spatial chirp, and

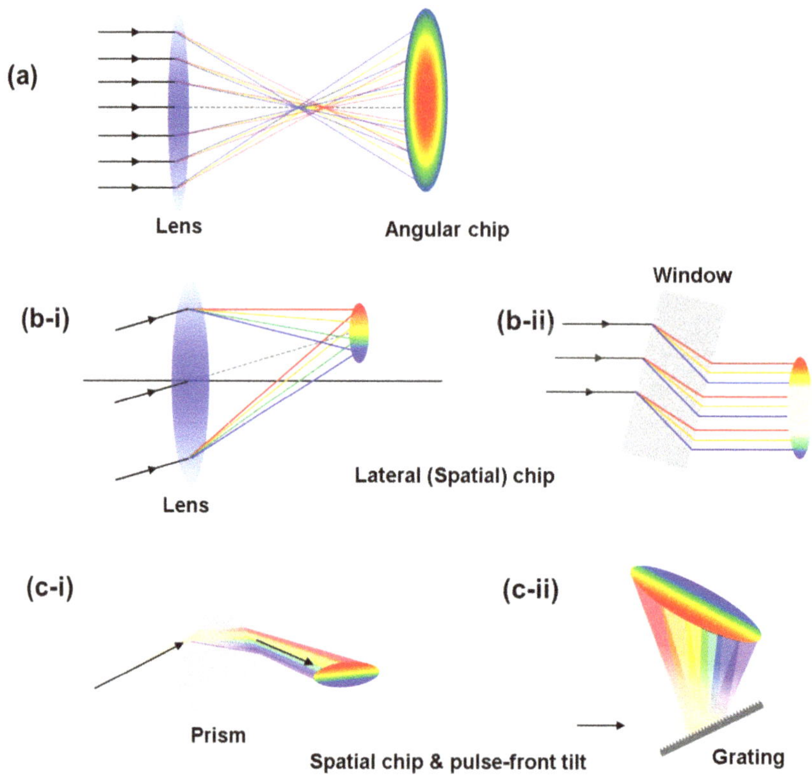


**Figure 2.16.** Chromatic aberrations of (a) angular chirp (or dispersion) by a lens, (b) lateral (or spatial) chirp by (i) a lens and (ii) a tilted window, and (c) spatial chirp with pulse-front-tilt by (i) a prism and (ii) a grating.

wavefront tilt as shown in figure 2.16 [48]. To avoid chromatic aberration, the lenses in the ultrashort laser beam lines should be achromatic lenses that fit to the bandwidth of the laser pulse. Prisms and wedged windows should not be used to avoid the chromatic aberrations, and the laser beam should be normally incident to the windows. Even with careful cautions on transmissive optics, a slight misalignment of pulse compressor can cause non-negligible chromatic aberrations.

The chromatic aberrations are not easy to measure and compensate properly. Here, an exemplary trick to compensate the chromatic aberration in an ultrashort high power laser system is as following: When the wavefront correction by the closed loop compensation between wavefront sensor and deformable mirror is not converging, or focal spot is ugly after the wavefront correction, the chromatic aberrations should be considered for the reason of the wavefront correction malfunction. Then, narrow bandpass filters with same frequency band can be applied to the wavefront sensor and the focal spot measurement. After the wavefront correction with a band-pass filter, the focal spot with the same bandpass filter should show better focal quality. If the focal spot is elongated or distorted by removing the bandpass filter in focal spot measurement, it is highly possible that the chromatic

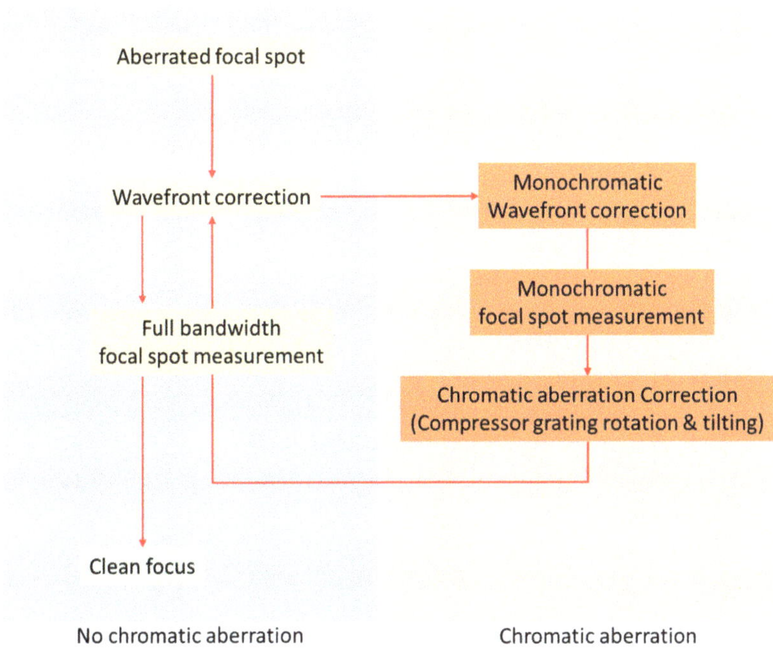


**Figure 2.17.** Workflow for compensating aberrations of an ultrashort ultra-intense laser pulse.

aberrations cause the degradation of the focusability. If there is chromatic aberration, it should be checked whether there are improper transmissive optics in the laser beam lines. If there are no such improper transmissive optics, then mostly misalignment of the gratings in pulse compressor can cause the chromatic aberration. If the full-band focal spot measurement appears elongated horizontally, then very slight rotation of the last grating in the compressor can compensate the chromatic aberration. The focal spot is elongated and tilted, then the last grating should be tilted and rotated. By repeating the wavefront correction and grating angle detuning, the aberrations of laser pulse can be compensated, and a clean focal spot can be obtained (figure 2.17).

## Exercises

(1) What is the chirp-free pulse duration of Gaussian shape pulse having 45 nm bandwidth in full width at half maximum (FWHM) at 800 nm central wavelength?
(2) What will be the pulse duration of chirped Gaussian pulse having GDD of 400 $fs^2$, 45 nm bandwidth in FWHM at 800 nm central wavelength?
(3) What will be the estimate of the duration of $sech^2$-shaped pulse from a measured intensity autocorrelator signal of 200 fs in FWHM?
(4) Draw interferometric autocorrelator signal for a chirp-free 50 fs Gaussian pulse at 800 nm central wavelength.

(5) Draw FROG signal for a chirp-free 50 fs Gaussian pulse at 800 nm central wavelength.
(6) Draw FROG signal for chirped Gaussian pulse having GDD of 400 fs$^2$, 45 nm bandwidth in FWHM at 800 nm central wavelength.
(7) What is Rayleigh length for a Gaussian laser beam having a waist size of 50 μm and a 532 nm wavelength?
(8) What will be the $M^2$ factor of a laser beam at 532 nm wavelength having a beam size of 200 μm after propagating 30 mm from the beam waist and beam waist size of 50 μm?
(9) What will be the focused beam size in FWHM for a Gaussian beam with 800 nm wavelength and diameter of 30 mm in FWHM focused by an OAP with focal length of 65 mm?
(10) Calculate laser intensity when a Gaussian laser pulse with pulse duration of 25 fs is focused to 20 μm in FWHM.
(11) If you have elliptical focal spot, what are possible reasons for elongation of the focal spot?

## References

[1] McClung F J and Hellwarth R W 1962 Giant optical pulsations from Ruby *J. Appl. Phys.* **33** 828
[2] Taylor N 2007 *Laser: the Inventor, the Nobel Laureate, and the Thirty-year Patent War* (Bloomington, IN: Iuniverse Inc)
[3] Hargrove L E, Fork R L and Pollack M A 1964 Locking of He–Ne laser modes induced by synchronous intracavity modulation *Appl. Phys. Lett.* **5** 4–5
[4] Strickland D and Mourou G 1985 Compression of amplified chirped optical pulses *Opt. Commun.* **56** 219
[5] Mourou G A, Tajima T and Bulanov S V 2006 Optics in the relativistic regime *Rev. Mod. Phys.* **78** 309
[6] Yoon J W *et al* 2021 Realization of laser intensity over $10^{23}$ W/cm$^2$ *Optica* **8** 630–5
[7] Gibbon P 2005 *Short Pulse Laser Interactions with Matter: an Introduction* (London: Imperial College Press)
[8] Umstadter D 2003 Relativistic laser–plasma interactions *J. Phys. D: Appl. Phys.* **36** R151
[9] Danson C N *et al* 2019 Petawatt and exawatt class lasers worldwide *High Power Laser Science and Engineering* **7** E54
[10] Hargrove L E, Fork R L and Pollack M A 1964 Locking of He–Ne laser modes induced by synchronous intracavity modulation *Appl. Phys. Lett.* **5** 4
[11] Lamb W E 1964 Theory of an optical maser *Phys. Rev.* **134** A1429
[12] Lin W, Lai T, Cheng Y, Zheng X, Xu X and Mo D 1991 60-fsec pulse generation from a self-mode-locked Ti:Sapphire laser *Opt. Lett.* **16** 42–4
[13] Squier J, Salin F and Piché M 1991 Mode locking of Ti:$Al_2O_3$ lasers and self-focusing: a Gaussian approximation *Opt. Lett.* **16** 1674–6
[14] Wang J and Chen S 1991 Self-starting issues of passive self-focusing mode locking *Opt. Lett.* **16** 1689–91
[15] Spielmann C, Brabec T, Krausz F and Curley P F 1992 Kerr lens mode locking *Opt. Lett.* **17** 1292–4

[16] Boyd R W 2020 *Nonlinear Optics* 4th edn (New York: Academic)
[17] Kerr J 2009 LIV. A new relation between electricity and light: dielectrified media birefringent (second paper) *London, Edinburgh Dublin Phil. Mag. J. Sci.* **50** 446
[18] Kerr J 2009 XL. A new relation between electricity and light: dielectrified media birefringent *London, Edinburgh Dublin Phil. Mag. J. Sci.* **50** 337–48
[19] Weinberger P 2008 John Kerr and his effects found in 1877 and 1878 *Phil. Mag. Lett.* **88** 897–907
[20] Sung J H, Lee S K, Yu T J, Jeong T M and Lee J 2010 0.1 Hz 1.0 PW Ti:Sapphire laser *Opt. Lett.* **35** 3021
[21] Yu T J, Lee S K, Sung J H, Yoon J W, Jeong T M and Lee J 2012 Generation of high-contrast, 30 fs, 1.5 PW laser pulses from chirped-pulse amplification Ti:sapphire laser *Opt. Express* **20** 10807
[22] Sung J H, Lee H W, Yoo J Y, Yoon J W, Lee C W, Yang J M, Son Y J, Jang Y H, Lee S K and Nam C H 2017 4.2 PW, 20 fs Ti:Sapphire laser at 0.1 Hz *Opt. Lett.* **42** 2058
[23] Armstrong J A 1967 Measurement of picosecond laser pulse widths *Appl. Phys. Lett.* **10** 16
[24] Sala K L, Kenny-Wallace G A and Hall G E 1980 CW autocorrelation measurements of picosecond laser pulses *IEEE J. Quantum Electron.* **16** 990
[25] Hirayama T and Sheik-Bahae M 2002 Real-time chirp diagnostic for ultrashort laser pulses *Opt. Lett.* **27** 860–2
[26] Diels J C and Rudolph W 2006 *Ultrashort Laser Pulse Phenomena: Fundamentals, Techniques, and Applications on a Femtosecond Time Scale (Optics & Photonics Series)* (Cambridge, MA: Academic Press) 2nd edn
[27] Kane D J and Trebino R 1993 Single-shot measurement of the intensity and phase of an arbitrary ultrashort pulse by using frequency-resolved optical gating *Opt. Lett.* **18** 823–5
[28] Kane D J and Trebino R 1993 Characterization of arbitrary femtosecond pulses using frequency-resolved optical gating *IEEE J. Quantum Electron.* **29** 571
[29] Walmsley I A and Wong V 1994 Analysis of ultrashort pulse-shape measurement using linear interferometers *Opt. Lett.* **19** 287–9
[30] Gallmann L, Sutter D H, Matuschek N, Steinmeyer G and Keller U 2000 Techniques for the characterization of Sub-10-Fs optical pulses: a comparison *Appl. Phys. B Lasers Opt.* **70** S67
[31] Walmsley I A and Iaconis C 1998 Spectral phase interferometry for direct electric-field reconstruction of ultrashort optical pulses *Opt. Lett.* **23** 792–4
[32] Tournois P 1997 Acousto-optic programmable dispersive filter for adaptive compensation of group delay time dispersion in laser systems *Opt. Commun.* **140** 245
[33] Chang I C 1992 Collinear beam acousto-optic tunable filters *Electron. Lett.* **28** 1255
[34] Kim H T *et al* 2017 Stable multi-GeV electron accelerator driven by waveform-controlled PW laser pulses *Sci. Rep.* **7** 1
[35] Siegman A E 1986 *Lasers* (Melville, NY: University Science Books)
[36] Gouy L G Sur Une Propriété Nouvelle Des Ondes Lumineuses (https://worldcat.org/ko/title/sur-une-propriete-nouvelle-des-ondes-lumineuses/oclc/25595928).
[37] *ISO 11146-2:2021—Lasers and Laser-Related Equipment—Test Methods for Laser Beam Widths, Divergence Angles and Beam Propagation Ratios—Part 2: General Astigmatic Beams* (https://iso.org/standard/77770.html).
[38] Siegman A E 1993 Defining, measuring, and optimizing laser beam quality *Proc. SPIE* **1868** 2

[39] Zernike von F 1934 Beugungstheorie des schneidenver-fahrens und seiner verbesserten form, der phasenkontrastmethode *Physica* **1** 689
[40] Niu K and Tian C 2022 Zernike polynomials and their applications *J. Opt.* **24** 123001
[41] Coronel M, Carvajal R, Escárate P and Agüero J C 2021 Disturbance modelling for minimum variance control in adaptive optics systems using wavefront sensor sampled-data *Sensors* **21** 3054
[42] Hartmann J 1904 Objektiv untersuchungen *Zeitschrift Für Instrumentenkunde. Berlin Verlag von Jul. Springer.* **24** 1
[43] Shack R V and Smith F D 1971 Production and use of a lenticular Hartmann screen *J. Opt. Soc. Am.* **61** 656
[44] QWLSI Wavefront Sensing Technology—Phasics (https://phasics.com/en/company/unique-wavefront-sensing-technology/)
[45] Babcock H W 1953 The possibility of compensating astronomical seeing on JSTOR *Publ. Astron. Soc. Pac.* **65** 229
[46] Madec P-Y 2012 *Overview of deformable mirror technologies for adaptive optics and astronomy* **8447** 22
[47] Srivastava V *et al* 2022 Piezo-deformable mirrors for active mode matching in advanced LIGO *Opt. Express* **30** 10491–501
[48] *Spatio-Temporal Distortions* (https://swampoptics.com/spatio-temporal-distortions.html)

# Chapter 3

# Instrumentations for laser–plasma interaction with ultra-intense lasers

Experiments with laser plasma require a driving laser, a target system, and diagnostics. In chapter 2, the fundamentals of ultra-intense laser technologies are discussed. In this chapter, instrumental techniques other than laser technology for laser–plasma interactions will be described. Even though experimental setups vary, this chapter will introduce common technologies pertaining to vacuum, focusing, the target, and the detector.

## 3.1 Vacuum systems

As described in the preceding chapter, most ultra-intense lasers utilize chirped pulse amplification technology. When the compressed pulse, which typically has energy per pulse greater than tens of mJ, propagates in air, the Kerr effects of the air medium can deform the laser pulse. Thus, the pulse compressor, beam transport, and target chamber should be maintained in a vacuum. In this section, the fundamental concept of vacuum systems for ultra-intense laser–plasma interaction experiments is presented.

### 3.1.1 Vacuum pumps and gauges

A vacuum is a relative state in which the chamber pressure is less than that of the surrounding atmosphere. A vacuum pump is a piece of apparatus that can generate a partial or low-pressure vacuum by expelling gas or air molecules from a sealed chamber. There are numerous types of vacuum pumps, which can be categorized by the mechanisms of vacuum formation: positive displacement (PD), momentum transfer (MT), and entrapment [1–4].

PD pumps are appropriate for low or middle vacuums in which the remaining gas behaves as a fluid. PD pumps are operated by a mechanically moving component that cycles through intake, compression, and exhaust. PD pumps employ a variety of

doi:10.1088/978-0-7503-2703-9ch3

| Vacuum range | Absolute pressure (Torr) | Suitable pumps |
|---|---|---|
| Atmospheric | 760 | |
| Low vacuum | 1–760 | PD (piston, rotary vane, diaphragm, scroll, screw, and so on) |
| Middle vacuum | $10^{-3}$–1 | PD (scroll, screw, roots, and so on) |
| High vacuum | $10^{-8}$–$10^{-3}$ | MT (turbo pump, diffusion pump) |
| Ultra-high vacuum | $<10^{-8}$ | Entrapment (cryogenic, ion) |

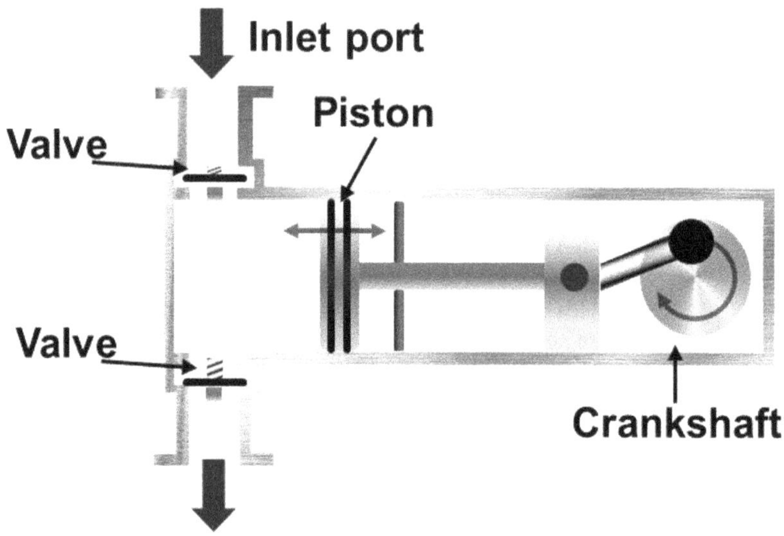


**Figure 3.1.** Example of PD pump (Piston pump).

mechanical methods, including pistons, screws, diaphragms, and vanes. Figure 3.1 shows the schematic drawing of piston pump. The PD pumps are utilized to form primary vacuum for MT pumps or to pump the foreline of an MT pump.

The fluid mechanisms utilized by PD pumps are inapplicable to high vacuum because the remaining gas lacks viscosity. The molecules of the residual gas behave as individual particles, hence the name molecular fluid. Therefore, each molecule's momentum must be transferred by dragging or pushing to create a high vacuum. Oil diffusion pumps and turbomolecular pumps are two common types of MT.

The oil diffusion pump [7] transfers momentum to the gas molecules using evaporated oil fluid. Figure 3.2 (a) shows the structure of an oil diffusion pump. An oil diffusion pump involves evaporating, ejecting, and condensing components. The oil is evaporated by a heater and ejected through nozzles on the top of vaporizing chamber. The vaporized oil exits the nozzles at supersonic speeds and drags gas

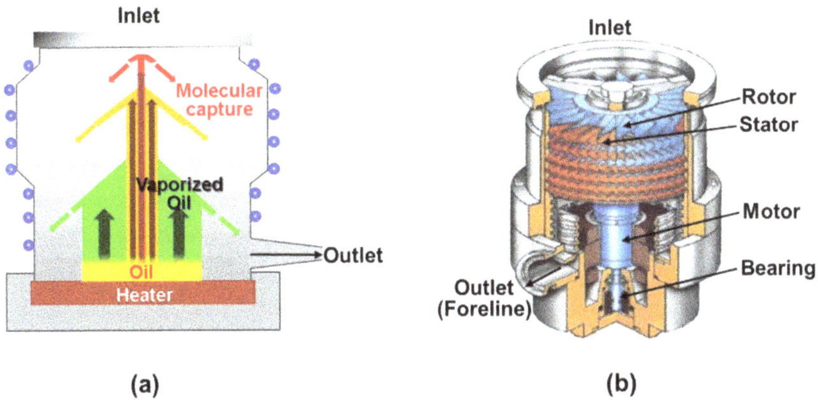


**Figure 3.2.** Schemes of typical momentum transfer pumps. (a) Oil diffusion pump and (b) turbomolecular pump. Reproduced from [8]. © The Author(s). Published by IOP Publishing Ltd. CC BY 4.0.

molecules down to the condensation part. In the condensing section, cooling coils capture the molecules dragged by the boiling oil, and the condensed oils return to the evaporating section. The collected gas molecules continue to flow towards the exhaust, while the pump's oils are repeatedly circulated to create a high vacuum.

The turbomolecular vacuum pump transfers momentum directly to molecules via rapidly rotating blades [8], as shown in figure 3.2 (b), whereas the oil diffusion pump utilizes oil conveyor. There are multiple stages of rotating and stationary turbine blades in turbomolecular pumps. The rotating and stationary blades are angled appropriately to transfer sufficient momentum to the gas molecules, transporting them to the subsequent stages until the exhaust. Since the rotational speed of the rotor must be very high in order to transport molecules with a very low mass, friction heat and rotor deflection limit are crucial design considerations for turbomolecular pumps.

Since MT pumps operate in the molecular fluid regime, their inlet parts must have at least a middle vacuum. Therefore, MT pumps are susceptible to failure when the inlet pressure is not at the middle vacuum level. Particularly when an inlet of a turbomolecular pump is exposed to a low vacuum, the fast-rotating blade can break and be ejected from the housing, causing severe injury to nearby persons. To avoid exposing the turbomolecular pump inlet to a low vacuum, the turbomolecular pump must be operated with caution and a safety device. PD pumps should be connected to the outlets of MT pumps to evaluate the accumulation of molecules at the outlets of MT pumps.

The oils from PD and MT pumps can cause optical degradation and damage by contaminating the surface of the optics when they are installed in a vacuum chamber. In vacuum chambers, therefore, oil-free PD pumps, so-called dry pumps such as scroll and screw pumps, along with turbomolecular pumps are mostly installed for ultra-intense laser interactions. However, these oil-free pumps are not resistant to corrosive or reactive chemicals. An oil diffusion pump with rotary vane

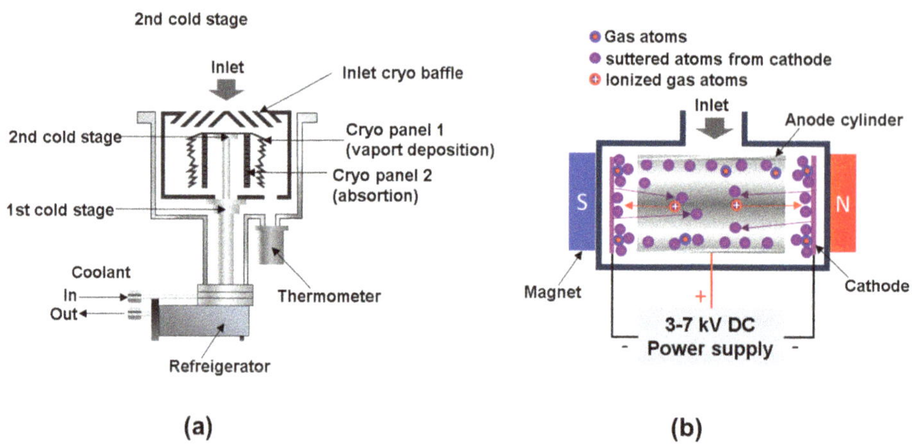


**Figure 3.3.** Schematics of (a) cryogenic vacuum pump and (b) ion vacuum pump.

pumps may be utilized as a suitable pumping system for corrosive or reactive chemicals.

Usually, a high vacuum is sufficient to prevent laser pulse distortion during transporting and focusing, whereas ultra-high vacuum (UHV) may be required for operating very high-voltage devices or transporting particles over long distances. For the formation of UHV, the molecules in a vacuum chamber must be reduced further by entrapment pumps below the level of vacuum created by MT pumps. One type of typical entrapment pump is a cryogenic pump [9], while the other is a sputter ion pump [10]. The cryogenic vacuum pump operates by cooling the gas to its condensation point. It captures gases like nitrogen and oxygen below 20 K, and it can condensate light gases like helium or hydrogen below 10 K. A typical cryogenic pump has a two-stage cooler, as shown in figure 3.3 (a); the first stage removes water vapor and oil by cooling to approximately 70 K–80 K, and the second stage removes gases by cooling to approximately 10 K–20 K. To capture the cooled gases, an adsorbent, such as activated charcoal, is added. Sputter ion vacuum pump (figure 3.3 (b)), also referred to as ion getter pump or ion pump, operates by ionizing incoming gases at the anode. The ionized gas then binds to a titanium cathode chemically or physically.

Vacuum gauges are indispensable instruments for measuring the vacuum level in vacuum chambers. There are two categories of vacuum gauges: direct and indirect gauges (figure 3.4) [11]. Direct vacuum gauges typically measure pressure by sensing the deformation of tubes, capsules, or diaphragms. Since direct gauges rely on the pressure exerted by gases, they primarily measure low and moderate vacuums in which the remaining gas behaves fluidly. Because pressure is only a function of density and temperature, and not molar mass, direct pressure gauges are independent of the gas species. The indirect gauges utilize density-dependent gas properties like thermal conductivity, ionization probability, and electrical conductivity. These properties depend on the molecular properties of the gases, so indirect gauges are dependent on the type of gas remaining.

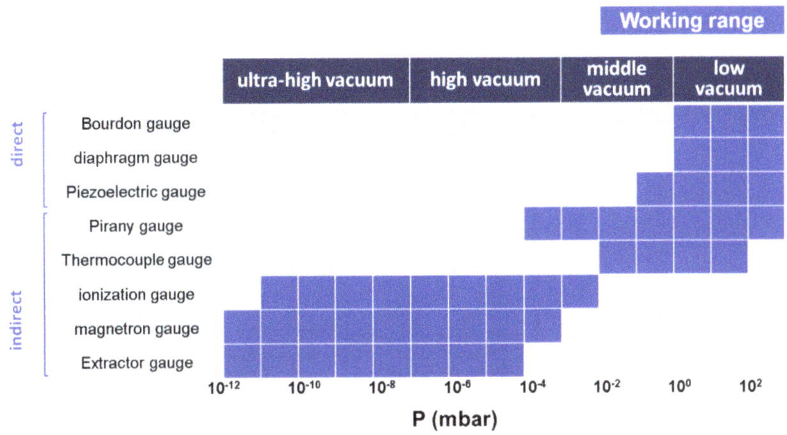


**Figure 3.4.** Various types of vacuum gauges and working pressures.

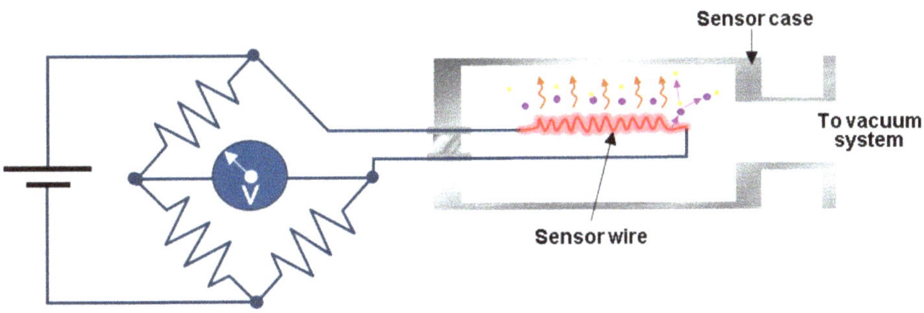


**Figure 3.5.** Diagram of Pirani gauge. Courtesy to sens4.com [12].

The most common indirect gauges are thermal conductivity gauges and ionization vacuum gauges. Thermal conductivity gauges utilize the pressure dependence of thermal conductivity in middle and high vacuum levels, whereas thermal conductivity has no pressure dependence near atmospheric pressure. The most well-known thermal conductivity gauge is the Pirani gauge (figure 3.5) [12]. A thin filament, heated to approximately 100° C–150° C, radiates and conducts the heat into the surrounding gas. When the pressure in a vacuum chamber falls to a few millibars, the mean free path of the gas is of the same order of magnitude as the diameter of the filament, and this type of heat transfer between the filament and surrounding gas becomes dependent on the pressure. The change in thermal conductivity is measured by setting the filament in the gauge head as a branch of Wheatstone bridge circuit.

Important tools for measuring gas pressures in the high and UHV regimes are ionization vacuum gauges [13, 14]. The ion gauge operates by ionizing gas molecules within the gauge chamber using an electric field. When electrons are driven in an electric field and obtain sufficient energy to form positive ions by colliding with gas

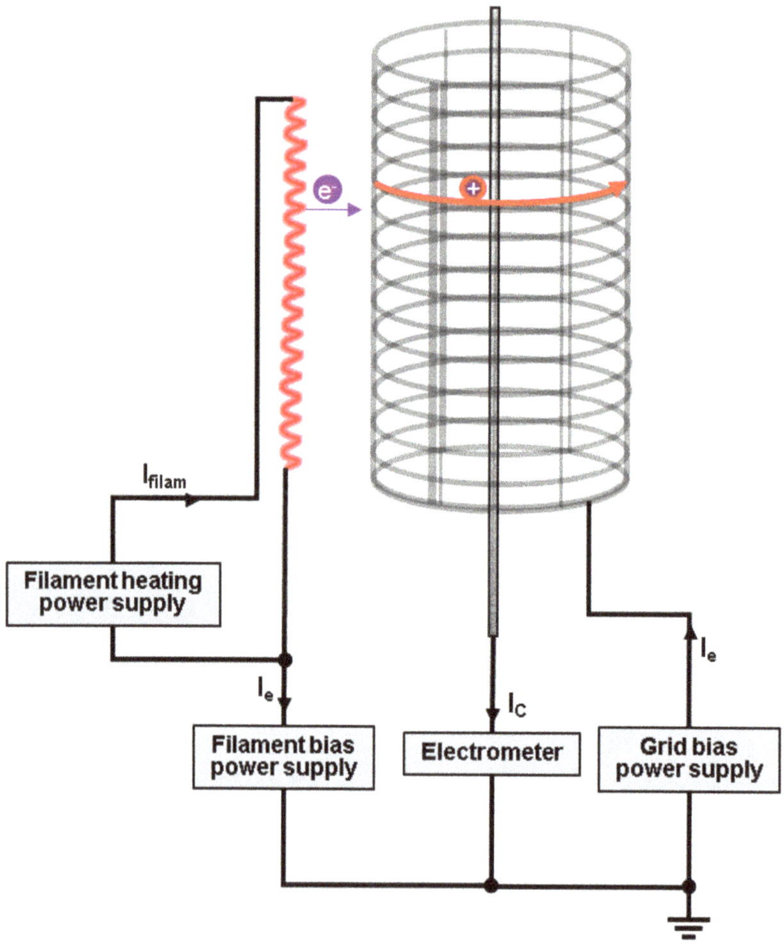


**Figure 3.6.** Structure of hot cathode ion gauge. Courtesy to Arun Microelectronics Ltd.

molecules, ionization occurs. The ions are then collected on a thin wire, known as the collector, and a current proportionate to the pressure is created. Since the ion yield is related to the pressure, the current is a measure of the pressure. Either a discharge at a high electric field strength, known as cold-cathode ion gauge, or the impact of electrons emitted from a hot cathode, known as hot-cathode gauge (figure 3.6), results in the creation of the ion current. One should be cautious when selecting and installing ion gauges, as their high voltage or light emission can be a noise source for highly sensitive sensors.

### 3.1.2 Vacuum techniques for laser–plasma experiments

In laser–plasma interaction experiments, the construction of a vacuum chamber is required for the pulse compressor, beam transport, and target installations. As stated in the previous subsection, vacuum pumps and gauges must be integrated into

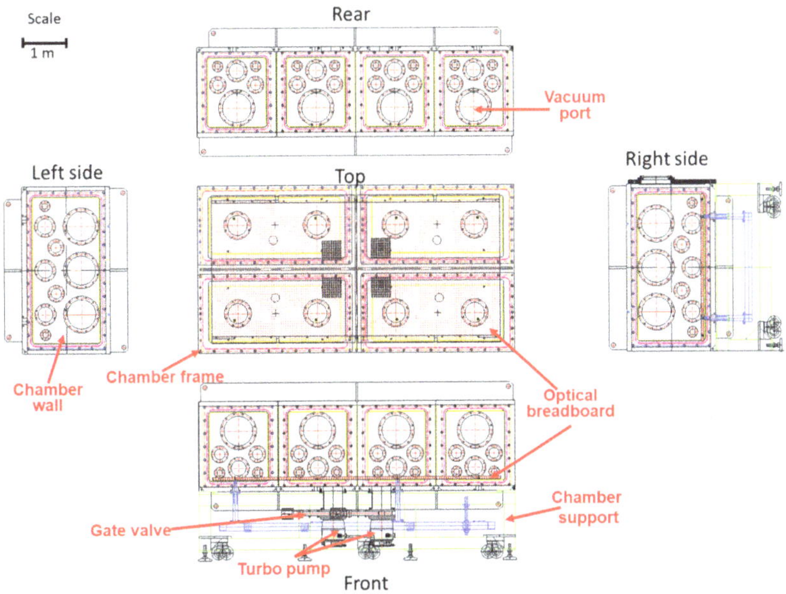


**Figure 3.7.** Exemplification of a vacuum chamber for laser–plasma interaction.

the vacuum chamber to achieve and confirm the desired vacuum level, while the design and construction of vacuum chambers continue to require specialized knowledge. In this subsection, we will discuss the most important techniques for building vacuum chamber.

The vacuum chamber consists of the chamber body, the sealing, and the flanges. The shape of the chamber body can be cylindrical, cuboid, or spherical. The chamber shapes should be chosen to accommodate the installed components. Figure 3.7 depicts a vacuum chamber of cuboidal shape. For the formation of a vacuum chamber, each surface component must be joined with sealing material or solder to create an airtight surface. A vacuum chamber should have ports for vacuum pump, vacuum gauge, beam input/output, electric signal lines, signal observation, etc. Quick flanges (QFs), also known as Kleinflansh; ISO vacuum flanges; and conflat (CF) flanges are the standards for vacuum ports [15]. QF is the O-ring-sealed, clamp-tight connection of a small-diameter tube to the chamber body. ISO vacuum flanges are used to connect large diameter port with O-rings and bolts. To create UHVs, CF-type flanges employ metal gaskets, primarily copper gasket. The combination of a metal gasket and a stainless-steel knife-edge flange makes for an almost impenetrable vacuum barrier. When an entire vacuum chamber is sealed with a metal gasket, outgassing and leakage can be reduced, allowing a UHV to be maintained. Thus, the flange sealing type should be carefully chosen with the required vacuum level for the experiment in mind.

The deformation of chamber shape by air pressure is another important consideration in vacuum chamber design for laser–plasma interactions. Every one $cm^2$ of the vacuum chamber surface is subjected to atmospheric pressure corresponding to one kg weigh. As the size of the vacuum chamber increases, air pressure becomes a significant factor. A 1 m × 0.5 m surface, for example, should withstand a weight of about five tons, which is heavier than the chamber wall itself and can cause deformation of the chamber shape. In this regard, spherical-shaped vacuum chambers have the advantage of air pressure forces valancing in all directions. The upper and bottom plates of a cylindrical chamber can be deformed, whereas all six sides of a cuboidal chamber can be deformed. In contrast, it is difficult to install components within a spherical chamber, so spherical chambers are mostly used for experiments involving many laser inputs to the chamber's center, such as laser fusion. Cylindrical chambers are useful for installing a target in the center and an off-axis parabolic (OAP) mirror with a short focal length, whereas cuboid chambers are preferred for long-focal-length focusing mirrors and inline detection arrangements. Strong support structures are required for cylindrical or cuboidal chambers to prevent deformation of the chamber surface. The installation of an optical breadboard in the target chamber or compressor chamber facilitates the installation of optical components within the chamber. Since air pressure will inevitably deform the chamber wall, it is preferable to separate the optical breadboard from the chamber wall. As illustrated in figure 3.7, the optical breadboard is not attached to the chamber wall, but rather to the chamber bottom support. The connection between the optical breadboard and the bottom support is sealed by the bellow flange of the chamber's underside. Consequently, the air pressure on the surface of the vacuum chamber and optical arrangement in the chamber are the most crucial factors in determining chamber structure.

## 3.2 Optical systems

### 3.2.1 Selections of optics

Optics of various types are required for laser–plasma experiments. Optics for femtosecond (fs) high-power lasers after the pulse compressor will be introduced here. After the compressor, refraction optics such as lens and prism are generally avoided because dispersions and nonlinear effects of the transmitting optics can distort the intense laser pulse. Beam transport and target chambers therefore employ reflection optics, predominantly mirrors. To choose mirrors, one must consider coating, substrate, and shape. In this section, key considerations for selecting mirrors will be discussed.

Mirror coating is the primary consideration [16]. There are two different types of mirror coating: metallic coating and dielectric coating. Due to their reflectance over a very broad spectral range, from 450 nm to 20 $\mu$m, metallic-coated mirrors are suitable for general applications. They are also insensitive to polarization and incidence angle. However, their soft coating surface makes them more prone to damage, so special care must be taken when cleaning them. In addition, the maximum reflectance of a metallic mirror is approximately 95%, which can result in significant light loss by multiple reflections. A dielectric mirror is composed of multiple thin dielectic stacks with low and high refractive indexes, which induce constructive interference in the reflection direction of light reflected from different layers of the

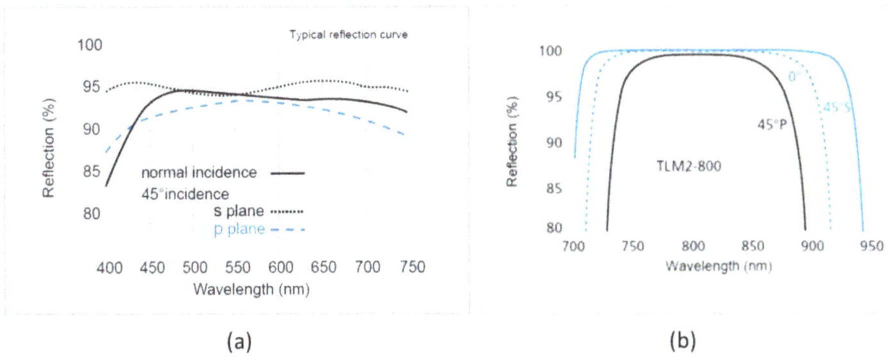


**Figure 3.8.** Reflectivity of (a) UV-enhanced Al mirror and (b) dielectric mirror for Ti:Sapphire fs laser. Courtesy to Korea Electro-Optics; reprinted with permission from [17].

dielectric stacks. Dielectric mirrors have higher reflectivity for a designed wavelength range than metallic mirrors. Figure 3.8 shows the different characteristics between metallic and dielectric mirrors. Dielectric mirrors have a durable coating that increases their resistance to laser damage and makes them easier to clean. The reflectance of dielectric mirrors is dependent on the incidence angle and polarization. The reflectivity of a dielectric mirror is close to 100 percent, and the loss of laser energy due to multiple reflections is negligible. In addition, when dielectric mirrors are utilized for ultrafast laser pulses, the dispersive effects of coating material and interface must be eliminated. For ultrafast laser pulses, a specially designed ultrafast laser mirror coating with negligible dispersion should be utilized.

For visible and near-infrared (NIR) optics, borosilicate, fused silica, and glass ceramic are the three most common substrate types [18–21]. Borosilicate glasses, such as Pyrex and BK7, are the least expensive option for the substrates of optics; however, they are not advised for temperature-sensitive applications, such as precision mirrors or highly thermally loaded mirrors. BK7 can be used for transmissive optics due to its low bubble and inclusion content, whereas Pyrex is not recommended for transmissive optics due to its inclusions and bubbles. Fused silica is an extremely pure synthetic amorphous silicon dioxide with a low inclusion content and a high refractive index homogeneity. The thermal expansion coefficient of fused silica glass is extremely low, and its optical qualities and ultraviolet (UV)-to-NIR transmittance are excellent. Transmissive and reflective optics utilize fused silica, especially when a high laser damage threshold is required. Glass ceramics are polycrystalline materials created by the controlled crystallization of base glass, resulting in a fine, uniform distribution of crystals throughout the bulk material. The glass ceramic optical material is typically the most expensive and qualified. Zerodur is a well-known glass ceramic optical substrate with a coefficient of thermal expansion close to zero and excellent homogeneity of this coefficient throughout the piece. Thus, Zerodur can be used for optical substrates that require extreme thermal stability, but due to inclusions, it should not be used for transmissive optics.

After the optical coating and substrate have been determined, the surface figure and quality must be specified to complete the optics selection. The qualification of surface figure is defined as the deviation from the designed shape in terms of the input light's wavelength ($\lambda$), which is typically 633 nm. Peak-to-valley or root-mean-square value can be used to describe the surface figure. A surface figure of $\lambda/10$ is typically acceptable for laser applications, but $\lambda/20$ or $\lambda/40$ may be required to obtain a high-quality focal spot with a short focal length focusing mirror. The surface quality of an optical component is an assessment of surface flaws, such as scratches and digs. The surface quality is more important for laser applications than for imaging applications because surface imperfections can cause optical damage by intense laser pulses. Scratch-dig is a straightforward method for classifying the level of surface quality that was derived from a military standard (MIL-PRF-13830B). The term 'scratch' refers to an elongated mark on an optical surface. The scratch number is an approximation of the scratch's width, but it is measured as visibility or brightness using a particular illumination. A dig is a circular pit or mark on the optical surface. Dig is a bit more quantitative, as its diameter serves as a rough approximation. Lower numbers for scratch and dig indicate a higher-quality surface. If your product's optical surface has a scratch-dig number of 40–20, it is marginally acceptable for laser applications. The majority of laser optics have a surface quality of 20–10, whereas experiments requiring high precision under sensitive laser conditions require 10–5. The specified surface quality has a direct effect on cost; therefore, over-specification of your optics and employing a higher surface quality than necessary will result in unnecessary cost increases. However, substandard optical surface quality can result in a failed experiment. For an optical system to be cost-effective, it is crucial to comprehend the required specifications for optics.

### 3.2.2 Focusing geometry

In laser–plasma experiments, focusing optics are typically the final optics to the plasma target. Generally, there are two types of focusing optics: concave spherical mirror and off-axis parabolic (OAP) mirror [22]. Setting up a concave spherical mirror is the simplest way to focus ultra-intense lasers on a plasma target. Nevertheless, depending on the incidence angle and *F#*, a concave mirror can induce astigmatism and spherical aberration. *F#* represents the ratio between focal length and beam diameter. To minimize aberration, spherical mirrors with a focal length significantly longer than the beam's diameter, i.e. a large *F#*, are typically employed. If the laser beam size is small, oblique incidence of laser beam to spherical mirror, as shown in figure 3.9(a), can still be acceptable with a small amount of astigmatism; however, astigmatism cannot be eliminated in this configuration. As shown in figure 3.9(b), the spherical mirrors can be installed with a 45-degree holed reflection mirror and normal incidence to the focusing mirror. In this case, a small amount of central laser energy is lost, but the laser beam is not astigmatic. Also, spherical aberration can cause focusing issues for small *F#*. Therefore, concave spherical mirrors with a large focal volume and a large focal diameter are preferred

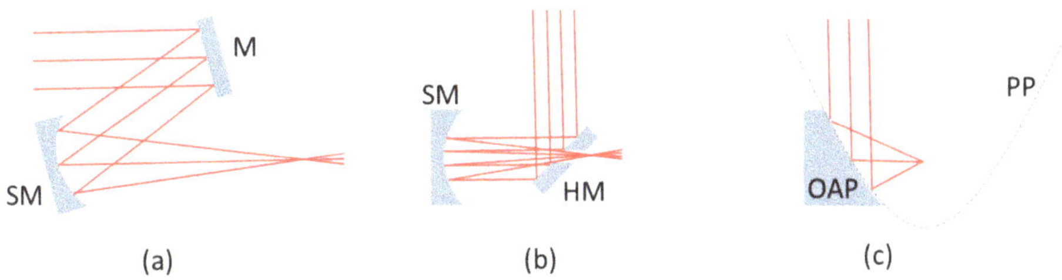


**Figure 3.9.** Optical payouts for (a) oblique incidence to spherical mirror, (b) normal incidence to spherical mirror with holed 45-degree flat mirror, (c) OAP mirror. M is flat mirror, SM is spherical mirror, HM is flat mirror with hole, and PP is parent parabolic mirror surface.

for large *F#*. Large *F#* will have a lower laser intensity for a given laser peak power than small *F#*.

An OAP mirror with a low *F#*, as depicted in figure 3.9(c), can be an excellent method for achieving a high laser intensity by forming a small laser focus. The OAP is a sectional cut from an off-axial portion of the 'parent' parabolic mirror, which is a larger parabolic mirror. OAP mirrors provide achromatic reflection, aberration-free focusing, and the ability to deviate the light path off axis for an incidence laser pulse to target [23]. If there are no aberrations in the wavefront of the incident laser pulse, the focal spot will have a diffraction-limited size, while improper alignment of OAPs can lead to extremely aberrated focus. Therefore, an OAP must be meticulously aligned using 6-axis fine adjustments.

### 3.2.3 Focal spot monitoring method

Laser focal spots, typically, are similar or smaller than a charge-coupled device (CCD) pixel. Therefore, an achromatic aspheric lens should be used to magnify the image of the focal spot. If high magnification is required, an achromatic micro-objective lens may be a suitable option. There are additional characteristics of magnification lenses that must be considered. The numerical aperture (NA) of magnifying optics should be larger than that of focusing optics, or the f-number of magnifying optics should be smaller than that of focusing optics. In optics, the NA of an optical system is a dimensionless number that describes the range of acceptable angles (figure 3.10). In microscopy, NA is typically used to describe the acceptance cone of an objective given by

$$\mathrm{NA} = n \sin(\theta) \tag{3.1}$$

where $n$ is the refraction index of the medium in which the lens is working (1.00 for air, and 1.33 for pure water), and $\theta$ is the half-angle of the maximum cone of light that can enter or exit the lens. For the maximum incidence angle, it is required that light should pass through the entire optical system, not just the entrance aperture. The f-number corresponds to approximately half of NA in air. If the NA of the objective lens is less than that of the focusing optics, the rays from the focusing lens cannot be collected by the objective lens, resulting in image

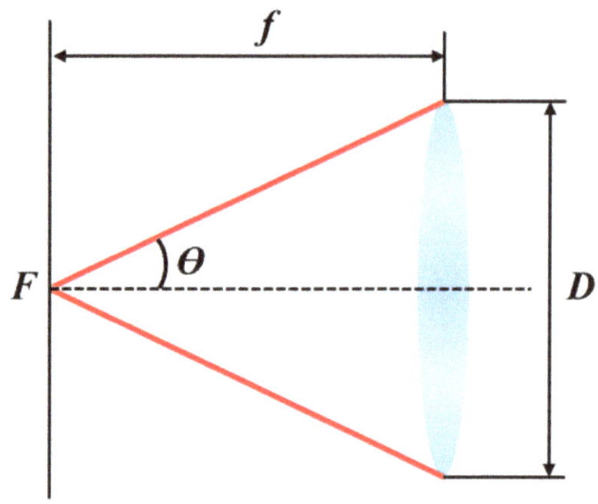


**Figure 3.10.** Simple geometry of a convex focusing lens.

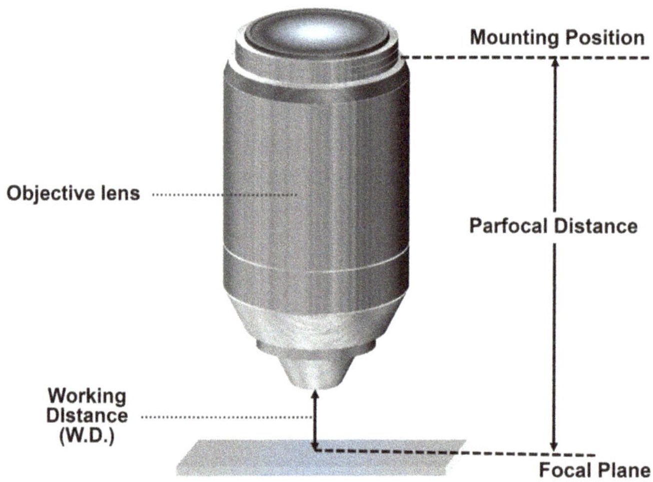


**Figure 3.11.** Schematic drawing of objective lens installation.

distortion of the focal spot. In consideration of the NAs of the focusing mirror and objective lens, care must be taken when selecting objective lens for magnification of focal spot image.

Another issue to consider is the working distance (WD) of an objective lens (figure 3.11). The WD is the distance between the front edge of the objective lens and the specimen surface [24]. The focal plane is the specimen position in the case of focal spot monitoring. If the WD is too small, the first optical surface of the objective lens will be exposed to extremely high laser intensity and will be easily damaged. Thus, the laser intensity at WD should be attenuated sufficiently to avoid nonlinear effects and lens damage in order to obtain a properly magnified focal spot image.

## 3.3 Target systems

Target conditions such as electron density, temperature, and atomic species are critical parameters for laser–plasma interactions [25]. The electron density of plasma, i.e. plasma density, is the most important parameter for determining the characteristic behavior of plasma in ultra-intense laser–plasma interaction. The plasma density can be classified by the critical density at which plasma frequency equals laser frequency. The formula for critical density is:

$$n_c = \frac{\varepsilon_0 m_e}{e^2}\omega_0^2 \approx \frac{1.22}{\lambda_0^2[\mu\text{m}]}[\text{in units of } 10^{21}\text{ cm}^{-3}] \tag{3.2}$$

where the $\varepsilon_0$ is vacuum permittivity, $m_e$ is electron mass, $e$ is electron charge, $\omega_0$ is angular frequency, and $\lambda_0$ is the wavelength of laser field. At the critical density, the refractive index of plasma becomes zero for the given laser frequency, causing the laser pulse to be reflected. Over-dense plasma occurs when the density of the plasma significantly exceeds the critical density, whereas under-dense plasma describes the opposite situation. When ionized by a laser field, gaseous and solid targets typically produce under-dense and over-dense plasmas, respectively. This section describes methods for creating gaseous and solid targets for producing under-dense and over-dense plasmas for laser–plasma experiments.

### 3.3.1 Low-density plasma mediums

Gaseous media are commonly used to produce low-density plasma. Figure 3.12 shows the three types of gas targets for laser-plasma interactions in under-dense conditions. Since ultra-intense laser–plasma experiments are conducted in vacuum chambers, the gases must be puffed into vacuum chambers or contained within a sealed volume. For puffing gases into vacuum target chambers, electric pulse values are typically used to reduce the amount of exhausted gas for maintaining the target chambers' vacuum level. There are numerous types of nozzle structures that can be used to create the desired gas profile and density at the interaction zone. The de Laval nozzle [26], which is a tube with a pinch in the middle, is a well-known

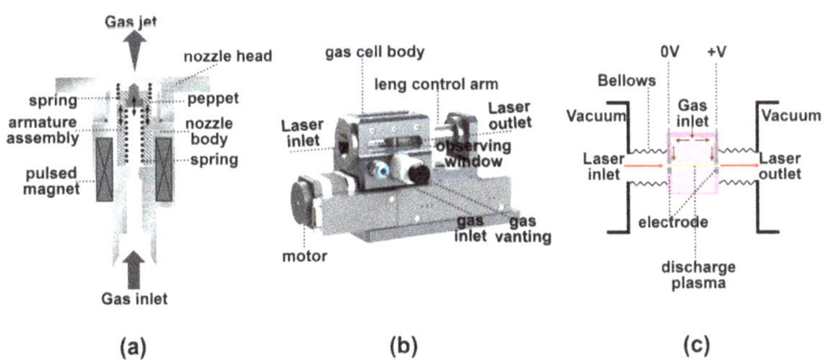


**Figure 3.12.** Three types of under-dense laser–plasma targets: (a) gas nozzle using electric pulse valve [30], (b) variable length gas cell (Reprinted with permission from [31]), and (c) capillary discharge plasma [32].

structure for forming a gas jet medium. By converting the flow's thermal energy to kinetic energy, de Laval nozzles can accelerate compressible fluid to supersonic speed to the thrust direction. Since the supersonic gas jets can provide lowly diverging gas flow, the laser–plasma interaction can occur away from the nozzle tip, which is advantageous for preventing nozzle damage and facilitating plasma diagnostics for the installation. However, the supersonic gas jet can induce shocks and turbulences, which can result in undesirable instabilities and lack of uniformity in the gas medium.

A gas cell is the most basic type of gas medium, containing gases in a sealed volume [27–29]. The gas cell should have entrance and exit pinholes to allow laser pulses to pass through the gas medium during ultra-intense laser–plasma experiments. The gas cells have benefits stable and uniform gas density distributions. Furthermore, by varying the distance from the entrance to the exit hole, the gas cell can provide variable length of the gas medium. It is not easy for gas jets and capillary discharge plasma medium to produce a medium with a length longer than 10 cm, but a gas cell can accomplish this feat by simply extending the distance between two pinholes. Gas cells typically provide a gaseous medium for lower electron density than gas nozzles because filling high-density gases in the gas cell produces a large amount of exhaust gas, which can reduce the target chamber's vacuum level.

Capillary discharge plasmas are formed by an electrical discharge through a narrow channel, or capillary. Initially the capillary is filled by neutral gas, typically hydrogen, and high voltage between electrodes ionizes the gas. Free-electron gas has refractive index lower than one, and the intense laser beam can be defocused by the higher electron density at the laser axis during the ionization of the neutral gases, which is often called plasma-defocusing effect. Due to the plasma-defocusing effect, there is a serious problem in intense laser pulse propagating for a long distance in a plasma medium. In contrast, the expanded plasma in the capillary, that has higher off-axial electron density than on-axis, can act as an optical waveguide like optical fibers. Ideally, a hyperbolic distribution of electron density and laser spot size corresponds to the matching condition of the plasma waveguide enables an intense laser pulse guiding flawlessly to a long distance. In capillary discharge channel, the laser alignment should precisely follow the preformed plasma channel direction to avoid wobbling of laser beam and damage of capillary wall.

### 3.3.2 High-density plasma mediums

Solid targets are the most common mediums for high-density plasma. Since an ultra-intense laser pulse causes damage to the target's irradiated area, the solid target should be displaced for each laser shot. The types of solid targets can be categorized by the thickness: bulk slab target with several-mm thickness, few-micron foil target, and ultra-thin sub-micron foil target (figure 3.13). The foil target can be made in the shape of a tape and mounted on a motorized tape holder for repeated irradiation. Typically, the ultra-thin foil targets are mounted on a plate with hole arrays.

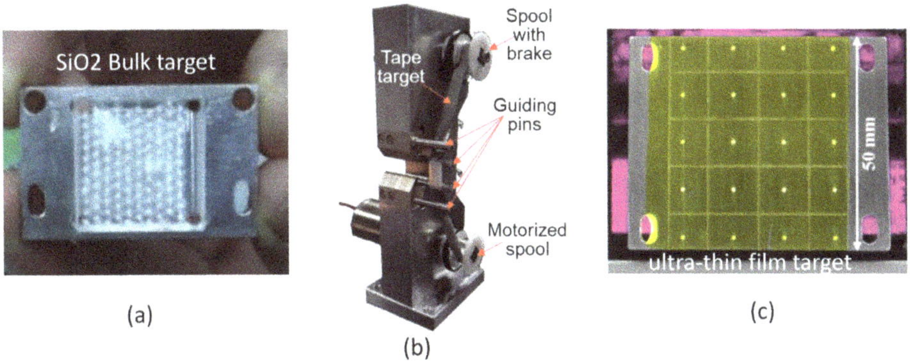


**Figure 3.13.** (a) Solid bulk target after laser irradiations [33], (b) tape target system (Reproduced from [34]. CC BY 4.0.), and (c) ultra-thin film target mounted on hole-array plate [35].

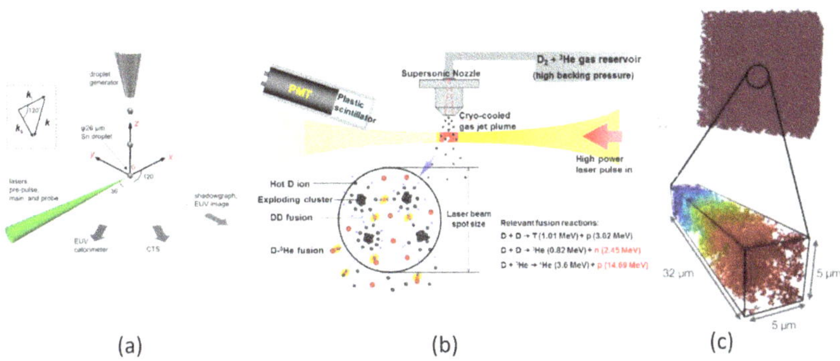


**Figure 3.14.** Examples of distinctive targets for high-density plasmas: (a) microdroplet target (Reproduced from [36]. CC BY 4.0.), (b) cluster target (Reprinted from [37], with the permission of AIP Publishing.), and (c) nano-structured target (Reproduced from [38]. CC BY 4.0.).

Therefore, target shapes and holders should be designed with the required target thickness in mind.

Liquid droplet targets, cluster targets, and nano-structured targets are distinct from conventional gas or solid targets (figure 3.14). The liquid droplet targets are created by puffing a small amount of liquid through a liquid nozzle into a vacuum chamber that has been shaped into a sphere by surface tension. The liquid droplet target with molten Sn has been utilized to generate a bright extreme-ultraviolet (EUV) source for EUV lithography. The cluster target consists of low-temperature gaseous atoms and molecules aggregated by the Van der Waals force. The cluster targets provide very local solid density with the average density of a gaseous medium, allowing ultra-intense lasers to efficiently produce hot dense plasma. The nano-structured targets are a type of solid target with nanometer-scale structures. By employing nano-structured targets, the absorption efficiency of ultra-intense lasers can be significantly increased.

## 3.4 Detection systems

### 3.4.1 Cameras

Cameras are two-dimensional image measurement devices for electromagnetic waves. For laser–plasma interactions, various types of cameras are used. The common types of scientific cameras used for laser–plasma interactions will be introduced in this section.

**(1) CCD and complementary metal-oxide semiconductor**

Charge-coupled devices (CCDs) and complementary metal-oxide semiconductor (CMOS) are the two most common types of cameras (figure 3.15) [39]. Both CCD and CMOS image sensors are made up of pixels that convert light into electrons by capturing light photons; the difference is in how the electron charges in each pixel are converted to image signal. Each pixel in a CCD sensor contains a potential well that collects photons and liberates electrons in proportion to the number of photons. The electrons in a pixel migrate down each row of the CCD, and the charge from each pixel is collected along the way. The charges reach an amplifier at the CCD's corner, where the charge corresponding to the photon number is converted into a voltage. The voltage signal is then digitalized using an analog-to-digital converter (ADC). In contrast, each pixel in a CMOS sensor contains several transistors that amplify the photo-induced charge and convert it to voltage. Each raw or column CMOS sensor includes an ADC to provide a digital image.

CCD devices, with specially designed manufacturing processes, can transport charges without distortion, resulting in high-quality, highly sensitive sensors with low read-out noise. CMOS chips have a simpler structure and can be manufactured using standard silicon semiconductor manufacturing processes, whereas CMOS sensors are less sensitive to incoming light and have higher noise than CCD sensors.

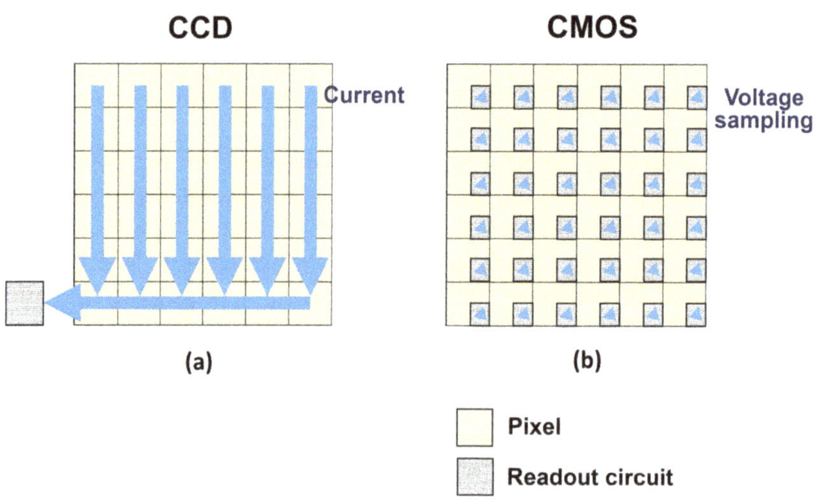


**Figure 3.15.** Schemes of CCD (left) and CMOS (right) camera.

CMOS sensors are much less expensive than CCD sensors, and their performance is rapidly improving. CCD sensors, on the other hand, are still required for some demanding applications, such as detecting very weak light signals, requiring very low noise, or detecting EUV/soft x-rays.

**(2) Front-illuminated and back-illuminated cameras**

A front illumination is a traditional method of producing digital cameras in which the wiring and circuitry are placed on the front surface to simplify the manufacturing process. However, the components on the chip's surface block a portion of the incoming light, so the photo-sensitive region can only receive the remainder of the incoming light, reducing the camera's efficiency. A back-illuminated camera [40, 41] locates the wires and circuitry elements behind the photo-sensitive region by flipping the silicon wafer and thinning the reverse side. Incoming light can strike the photo-sensitive layer without passing through the circuitry layer, increasing light-capturing efficiency from about 60% to more than 90%. Back-illumination technique is especially important for EUV/soft x-ray cameras because EUV/soft-ray lights are absorbed by most materials and the light must reach the photo-sensitive layer without passing through any intermediate layers. Figure 3.16 illustrates the structural distinctions between CMOS and CCD cameras. Due to additional manufacturing processes, back-illumination cameras are more expensive than front-illumination cameras, but they provide more sensitive responses to incoming light.

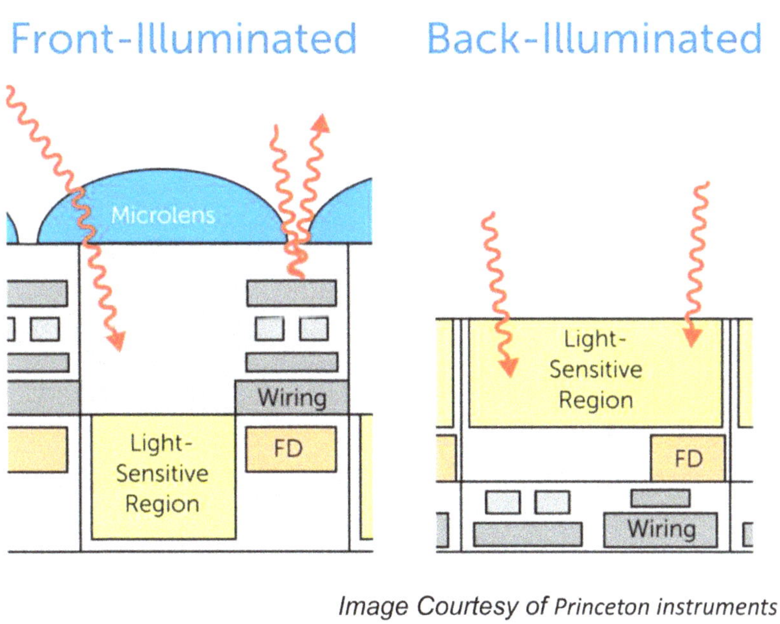


**Figure 3.16.** Structures of front-illumination (left) and back-illuminated (right) cameras. Reprinted with permission from [42].

**(3) Ultra-sensitive cameras**

Back-illumination cameras have a high sensitivity to incoming light, but it is difficult to detect extremely weak incoming light. Ultra-sensitive cameras that can amplify signals above the noise level should be used to detect extremely low fluence light. There are two types of ultra-sensitive cameras that are well developed: electron-magnifying CCD (EMCCD) [23] and intensified CCD (ICCD) (figure 3.17) [43]. EMCCDs are frame-transfer CCDs that have a signal-amplifying device before readout. Frame-transfer CCDs are divided into two sections: the image capture area, which captures the image, and the storage area, which stores the image before it is read out. EMCCD employs specialized extended serial registers, also known as multiplication registers, between the storage area and the readout amplifier. The multiplication registers greatly amplify photoelectron signals in silicon via impact-ionization processes with high clock voltages up to 50 V. Because this multiplication occurs before the readout amplifier, the signal is amplified before the read noise is applied. Thus, EMCCD cameras can detect single photons with sub-electron read noise using the photoelectron multiplication step.

ICCD cameras have a more intrinsic structure of amplification camera, by combining an external light intensifier and a CCD sensor. A photocathode, a microchannel plate (MCP), and a phosphor screen deposited onto a fiber optic make up the light intensifier. The photocathode converts incoming photons to electrons, and the MCP induces an avalanching process from each photoelectron into a cloud of electrons with a very high voltage up to several kV. The intensified photoelectron cloud by the MCP is converted to light by the phosphor screen on the front of the fiber optic, and the transferred light signal through fiberoptic bundle is detected by a CCD. The ICCD detector is highly sensitive, with single-photon detection capability, because incident photoelectrons are multiplied within the MCP. Furthermore,

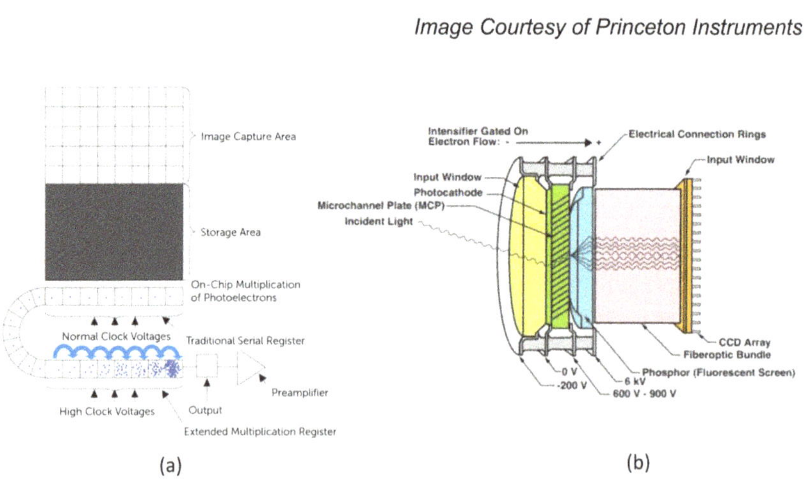


**Figure 3.17.** Schematic drawings of (a) EMCCD and (b) ICCD (courtesy of Princeton Instruments). Reprinted with permission from [43] and [44].

by controlling the voltage for the MCP, ICCD can provide very fast shutter or fast gating, allowing it to be used for applications requiring sub-nanosecond exposure time.

### 3.4.2 Spectrometers

Spectrometers are apparatuses that measure spectrums of lights. Most spectrometers use diffraction gratings whether they be transmission or reflection types (figure 3.18). A diffraction grating is an optical component with a periodic structure that diffracts light into different diffraction angles [45]. The diffraction angles are given by

$$m\lambda = d(\sin\alpha + \sin\beta), \tag{3.3}$$

where $m$ is diffraction order, $\lambda$ is wavelength, $d$ is the groove spacing on the grating, $\alpha$ is incidence angle, and $\beta$ is the diffraction angle. Thus, for a given incidence angle, diffraction order, and groove spacing, a wavelength of a light, i.e. spectrum, is resolved in space after a grating with a defined angle from the grating equation.

The Czerny–Turner spectrometer (figure 3.19) [45] is a common spectrometer design for UV, visible, and near-IR light. The light source (A) is incident on a normal-angle entrance slit (B), and the entrance slit trims the light in space. The light is collimated and incident to the diffraction grating by placing the slit at the effective focus of a curved mirror (the collimator, (C)). The collimated light is diffracted by the grating (D) and refocused on the camera (F) by another mirror (E). The slit image is related to the camera via the collimator and focusing mirrors, while the wavelength is resolved in the grating's dispersion direction. The spectrometer's spectral resolution is determined by the slit width, grating groove spacing, and optical resolution of the image relay optics. Nowadays, fiber-coupled spectrometers

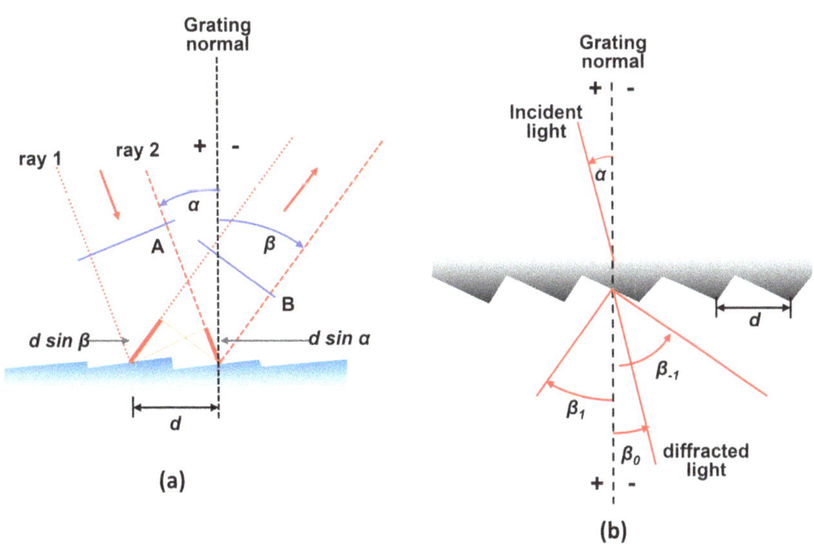


**Figure 3.18.** Diffraction of lights by (a) reflection grating and (b) transmission grating.

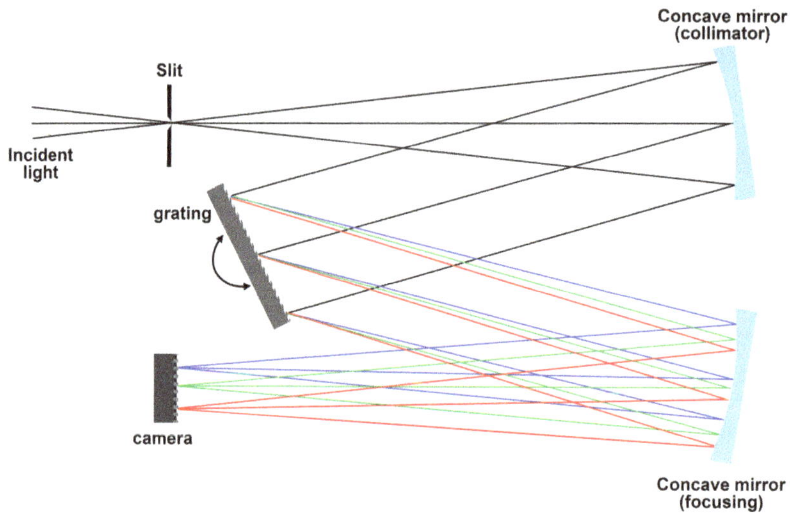


**Figure 3.19.** Scheme of Czerny–Turner spectrometer.

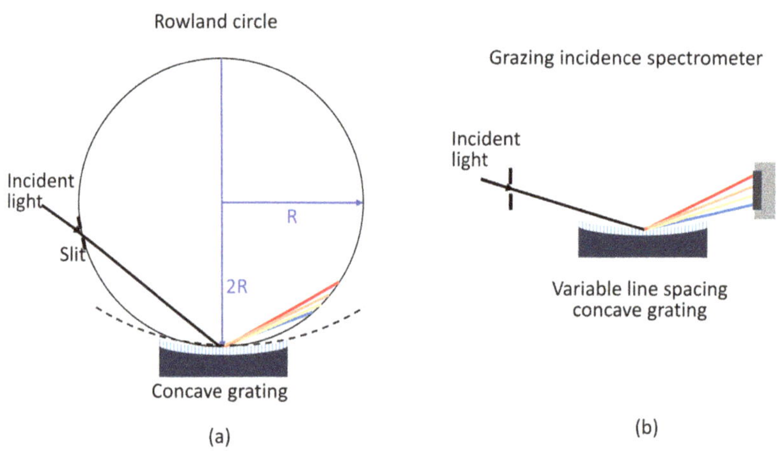


**Figure 3.20.** Layouts of (a) Rowland circle spectrometer and (b) grazing incidence spectrometer.

based on the Czerny–Turner spectrometer design, consisting of fiber input, image relay optics, diffraction grating, and line scanning sensor, are widely used to measure the spectrum of lights.

The Rowland circle spectrometer is the simplest spectrometer design (figure 3.20). The Rowland circle is a circle of radius $R$ tangent to a concave curved diffraction grating of radius $2R$, also known as a Rowland concave grating [46]. The spectrum of a light can be resolved by the Rowland concave grating, as its concave curved grating serves as both the diffraction grating and image relay optics. If the entrance slit of incoming light is positioned on the Rowland circle, the grating will focus reflected light at different points on the same circle depending on wavelength and diffraction order. Even though the Rowland circle spectrometer has a simple optical layout, aberrations of the optical layout restrict its applications. Particularly, the deformation of the spectral image is unavoidable on a camera with a flat sensor, as the spectral image is formed along the Rowland circle. However, the concave grating configuration is appropriate for grazing-incidence spectrometers for EUV or soft x-rays. EUV and soft x-ray sources are significantly absorbed by the most of materials, and the reflectivity of metal surfaces near normal incidence is extremely low. Thus, grazing-incidence optics, with an incidence angle close to parallel to the surface of the optics, are predominantly used for EUV and soft x-rays, except for multilayer mirrors designed for specific wavelengths. Consequently, EUV/soft x-ray spectrometers utilize the Rowland concave grating with a grazing-incidence angle, whereas flat-field configurations with variable line spacing concave gratings [47] are typically used to capture the spectral image without the typical deformation of the Rowland concave grating.

### 3.4.3 Interferometers

Interferometers are devices that extract information from the interference of coherent optical waves and are commonly used to measure refractive index distributions, material density distributions, surface shapes, and other parameters depending on optical path length or phase shift [45]. A single coherent light source is split into two beams that propagate in different optical paths in most interferometers. The two beams are combined to produce interference, and the interference fringes reveal the difference in optical path lengths or phase shift. Interferometers are frequently used in laser–plasma interactions to measure the gas density distributions of gaseous medium, or plasma density distributions for under-dense plasma or the peripheral part of over-dense plasma. Conventional Michelson type (figure 3.21 (a)) or Mach–Zehnder interferometers (figure 3.21 (b)) can be used for these purposes, while in-line interferometers, such as the Nomarski interferometer, are also frequently used to characterize plasma distributions.

The Nomarski interferometer (figure 3.21 (c)) is a simple in-line interferometer with no separate optical path [48, 49]. After passing through a specimen, a beam is divided into two beamlets using a biprism or a Wollaston prism. When using a

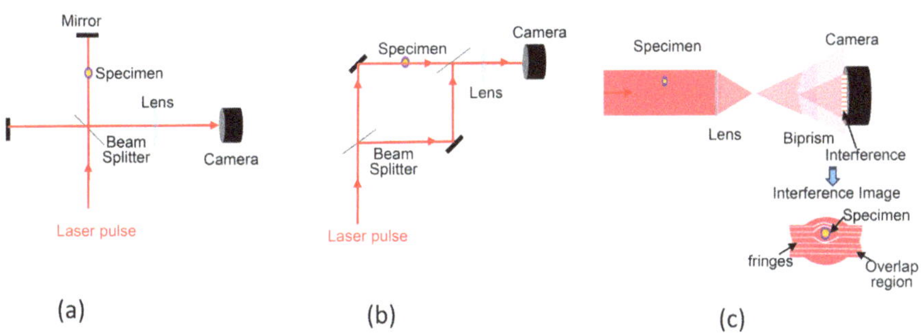


**Figure 3.21.** Optical layouts of (a) Michelson, (b) Mach–Zehnder, and (c) Nomarski interometers.

Wollaston prism, the unpolarized or diagonally polarized beam is split by the different refraction angles of the Wollaston prism for ordinary and extraordinary waves. After the diagonal polarizer, the diverging two beams overlap and form an interference pattern. In the case of biprism, the beam after a specimen spatially divided by the biprism and the overlapped part forms an interference pattern. The spatial division of a beam is known as the wavefront division method, whereas the beam division by beam splitter, which is used in Michelson and Mach–Zehnder interferometers, is known as the amplitude division method. The probe beam in Nomarski interferometers should be larger than twice the size of the specimen, and the unperturbed part of the beam serves as a reference wave.

For gas density or plasma density distributions, two-dimensional (2D) Fourier analysis is commonly used to extract information from the interference pattern. The procedure is similar to the spectral phase interferometry for direct electric-field reconstruction (SPIDER) reconstruction procedure described in section 2.3.1(3), but the dimension should be increased from one-dimensional to 2D. Figure 3.22 depicts an example of interferogram reconstruction. A Mach–Zehnder interferometer was used to capture the interferogram, with one path passing through an argon supersonic gas jet. The signal and background interferograms were captured with and without gas jet puffing. Interferograms were Fourier transformed, and the AC component was extracted. To obtain the phase distribution, the inverse Fourier transforms of the AC part of the signal and the background were performed, and the background phase was subtracted from the signal phase. Because the interferogram provided a lateral 2D projection of the gas jet distribution, the inverse Abel transform should be used to obtain the gas density distribution while assuming the gas distribution's cylindrical symmetry. The inverse Abel transform is given by

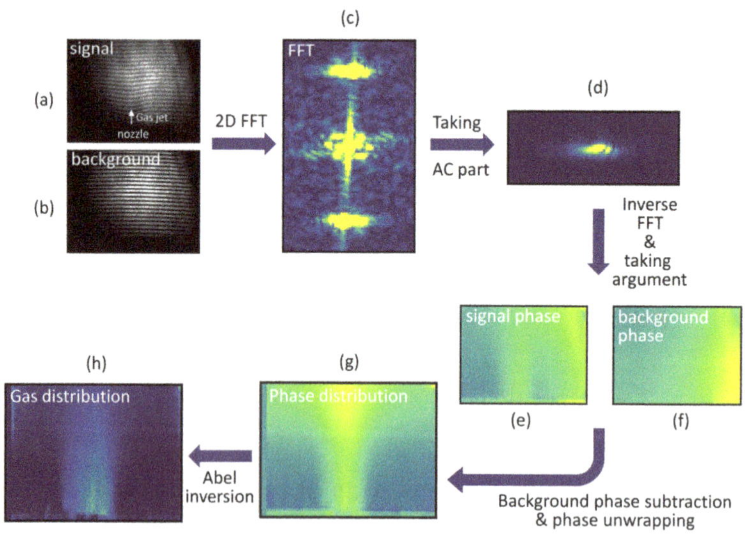


**Figure 3.22.** Reconstruction procedure of interferograms.

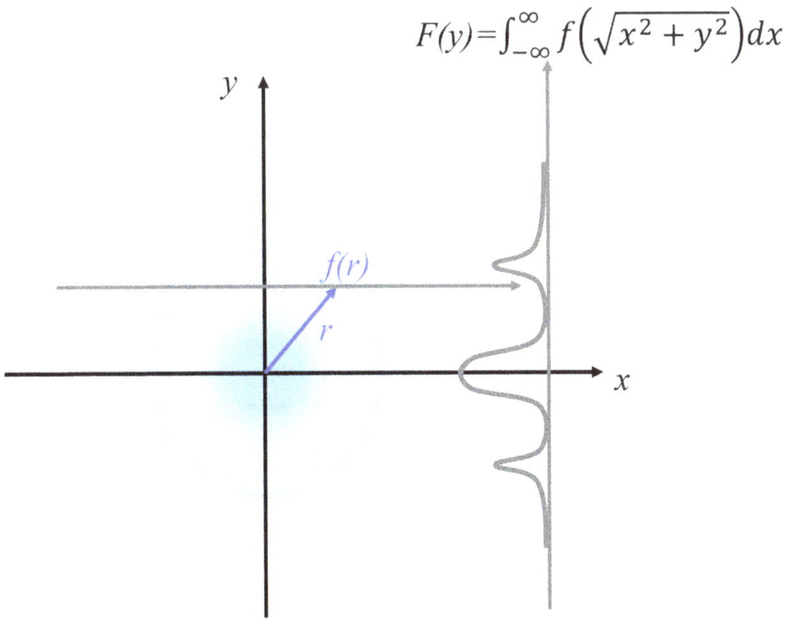


**Figure 3.23.** Geometry of Abel inverse transform.

$$f(r) = -\frac{1}{\pi}\int_r^{\infty} \frac{dF(y)}{dy}\frac{dy}{\sqrt{y^2 - r^2}} \tag{3.4}$$

where $f(r)$ is the symmetric function in $x$–$y$ plane and $F(x)$ is the laterally projected function along the $x$-axis in figure 3.23. The phase distribution function in figure 3.22 (g) is in the $y$–$z$ plane in figure 3.23, and the inverse Abel inversion function is in ($r$, $z$) coordinate. After converting the factor between phase and density for argon gas, the inverse Abel transform of the phase distribution yields the gas density distribution. As the example demonstrates, interferometers can be an essential tool for determining the characteristics of various phase objects.

### 3.4.4 Radiation detections

Laser–plasma interactions with ultra-intense lasers can produce nearly all spectral ranges of electromagnetic waves, from radio waves to gamma rays (figure 3.24). Electromagnetic waves in the infrared (IR), visible, and UV ranges can be detected by optical sensors, whereas microwaves and radio waves can be measured by conventional antennas. Most materials significantly absorb light with wavelengths less than 200 nm, which is known as vacuum UV (VUV) or EUV. As the wavelength decreases to 10 nm, light in this short wavelength range can ionize atoms from the bound state to the continuum state with a single photon, classifying this spectral range as soft x-rays. The grazing incidence spectrometer with a back-illuminated EUV/soft x-ray camera is a common detection method for EUV and soft x-rays. Hard x-rays are defined as photon energies greater than the K-shell ionization potential of most

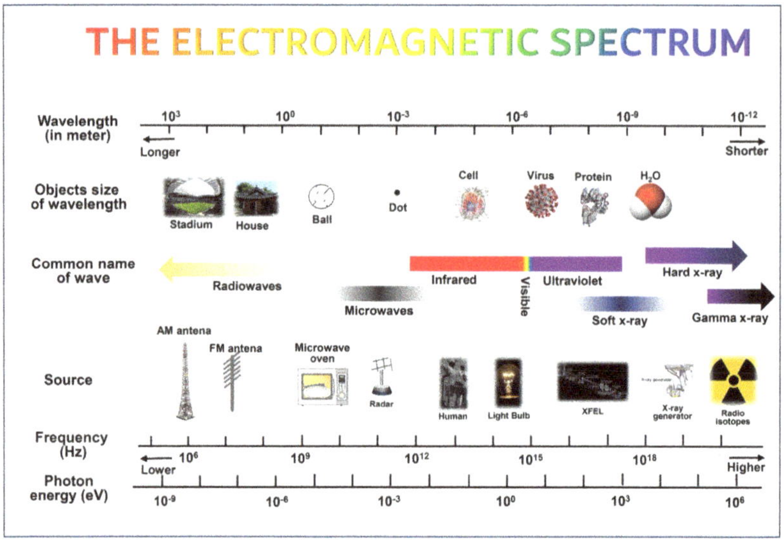


**Figure 3.24.** Spectrum chart of electromagnetic waves.

materials, with wavelengths typically less than 0.2 nm and photon energies greater than 5 keV. Gamma rays are defined by photon energy that can induce nuclear interactions, which corresponds to wavelengths less than 0.01 nm or photon energies greater than 100 keV. Hard x-rays and gamma rays penetrate almost all materials and are undetectable by standard optical sensors. Techniques for measuring hard x-rays and gamma rays from laser–plasma interactions with ultra-intense lasers are introduced in this section.

Continuous low-flux hard x-rays and gamma rays can be precisely measured using high-purity germanium detectors and multi-channel analyzers [50, 51]. The majority of the radiations produced by the interaction of an ultra-intense laser pulse and a plasma medium, on the other hand, are short bursts with a repetition rate of less than 10 Hz. As a result, traditional nuclear spectroscopy gamma-ray detectors are difficult to use with laser–plasma radiations. Low-energy hard x-rays of a few keV can be measured using concave crystals with grazing angle incidence geometry and hard x-ray cameras. It is, however, difficult to characterize radiations with energies greater than 10 keV in a single exposure. Figure 3.25 depicts several methods for measuring a single burst of hard x-rays or gamma rays from the laser–plasma interactions.

The simplest method is to use material absorption with various metals arranged transversely [52] or longitudinally [53]. Scintillators such as NaI(Tl), YAG(Ce), LYSO, and others can detect hard x-ray or gamma-ray radiations after the absorbing metal filters. Highly sensitive optical cameras can detect fluorescent signals. Another well-developed method for recording high-energy photons is photo-stimulated luminescence film, also known as 'image plate,' which was widely used in medical radiography. Even though image plates cannot be recorded in real time, they have the advantages of high sensitivity, high resolution, and a large detection area. The spectrum can be estimated indirectly by fitting the signal to calculated material responses for a given photon energy

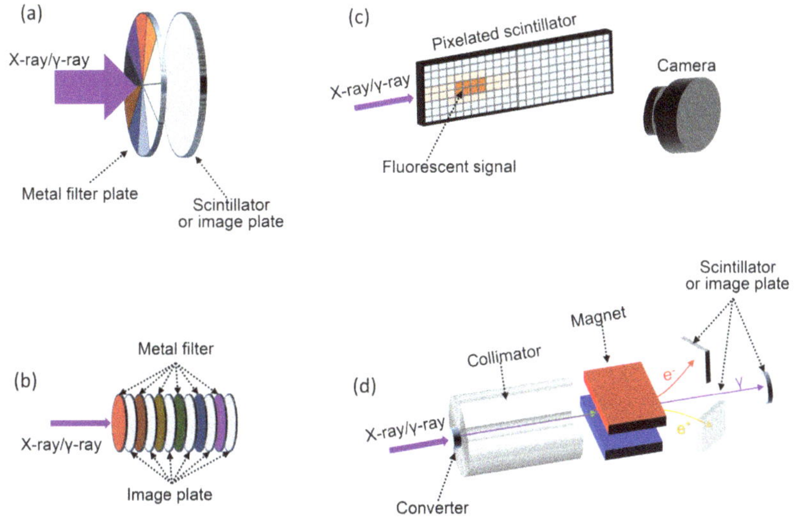


**Figure 3.25.** Schematic drawings of hard x-ray/gamma-ray spectroscopy by material absorption method using (a) angular filters, (b) longitudinal filters, and (c) pixelated scintillator. Panel (d) shows the scheme of Compton electron gamma-ray spectrometer.

while assuming the spectral profile shape. Monte Carlo simulation codes such as GEANT4 [54], FLUKA [55], and Monte Carlo N-Particle Transport (MCNPX) [56] can be used to calculate material responses. The pixelated scintillator method [57], which can combine the absorbing material and scintillator by arraying small pieces of scintillators and metal interlayers, is one of the advanced methods for measuring high-energy photons using material absorptions. The fluorescent pixels are separated by metal interlayers, and the radiation penetration length can be measured by observing the fluorescent signal at the side of the pixelated scintillator plate. Monte Carlo simulations for material responses and photon energy depositions can also be used to reconstruct the photon energy spectrum.

The spectrometry for high-energy photon using material absorption is easily constructed, although these methods have limitations of the spectral resolution and the accuracy of the reconstructed spectrum. Compton electron spectrometer [57] for high-energy high-flux gamma rays can be a solution to achieve high spectral resolutions. The Compton scattering between gamma photon and electrons in a converter material is given by

$$\Delta\lambda = \frac{h}{m_e c}(1 - \cos\theta), \tag{3.5}$$

where $\Delta\lambda$ is the change of the photon wavelength, $\lambda$ is Plank's constant, $m_e$ is electron mass, and $\theta$ is scattering angle. The energy transfer from high-energy photon to electron can be calculated directly using equation (3.5), and the electron energy by Compton scattering with low $Z$ materials is correlated to the incident gamma photon energy. Monte Carlo simulations can also calculate the correspondence between incident photon energy and Compton electron. To avoid strong multiple scatterings and pair creations by the Bethe–Heitler process, the convert material should be low-$Z$ materials with optimized thickness to the incident photon energy. The background signal caused by pair creations can be subtracted by measuring the electron and positron signals simultaneously after a dipole magnet, but a large amount of pair creation can prevent proper measurement of gamma photon energy. For the converters, low-$Z$ materials with a minimum thickness for Compton scatterings are recommended.

### 3.4.5 Charged particle detections

The interactions of ultra-intense lasers with plasmas can accelerate charged particles, especially for relativistic laser–plasma interactions. The energy of charged particle can be resolved by applying static magnetic or electric fields and using particle-sensing technologies. Figure 3.26 shows the typical relativistic electron measurement setup. The electrons in a magnetic field are deflected and the electrons are dispersed in orthogonal direction to particle propagation and magnetic field directions. For the scintillator screen, LANEX films produced by Kodak are commonly used. The accelerated electron beams by laser wakefield acceleration described in chapter 4 often have the problem of unstable beam pointing. A beam collimator in front of dipole magnet or multiple fluorescent plates are utilized to obtain the correct electron energy. The multiple-screen method is not proper for low-energy electron

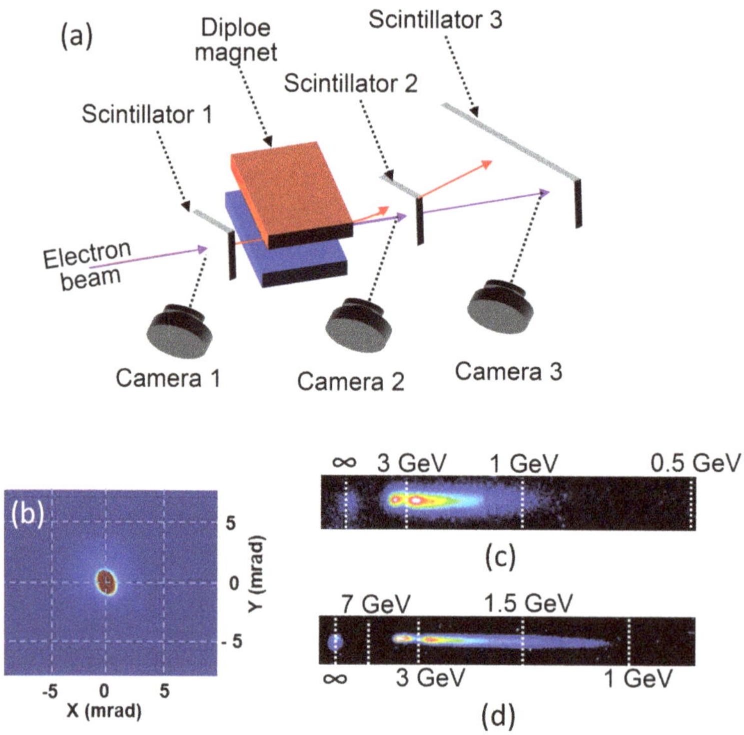


**Figure 3.26.** (a) Layout of dipole magnet electron spectrometer, (b) a measured electron beam profile in front of the dipole magnet, (c) an electron spectrum near the end of the dipole magnet, and (d) the other electron spectrum with a distance from the dipole magnet.

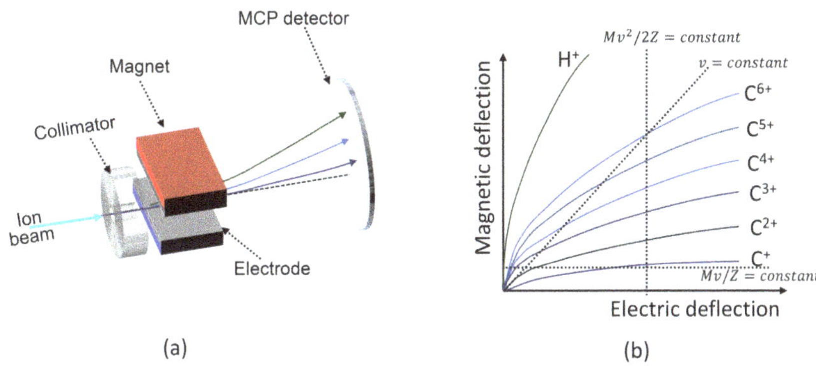


**Figure 3.27.** Drawing of (a) Thomson parabola ion spectrometer and (b) an example of typical ion traces by the Thomson parabola.

below 100 MeV due to the scattering of the electron beam by the screens, while multiple-screen method can be suitable for high-energy electrons over few 100 MeV energy. The recalibration of the correct electron beam energy can be done by the particle tracking method for given electron energy and incidence angle to the dipole magnet. The details of the recalibration method are described in reference [58]. It should be noted that the fringe field, stray fields at the edge of magnets or electrodes, should be considered to calculate the energy of charged particle measured by dipole-magnet electron spectrometer or Thomson parabola ion spectrometer.

The protons and ions from the laser–plasma interactions can be resolved by applying electric field and magnetic field simultaneously, called the Thomson parabola ion spectrometer. The Thomson parabola consists of collimator, dipole magnet, electrodes, and ion detector (figure 3.27) [58]. The strengths of magnetic field and electric field can be determined by considering ion energies and ion species. For ion detection in real time, a multi-channel plate with a phosphor screen and a camera can be used. The ion can be detected by Columbia Resin No. 39 (CR-39) [58] or radio-chromic films [58], while these detectors can be used for single record and should be processed after irradiations. Especially, the CR-39 can provide information on the flux of produced ion precisely. Figure 3.25(b) shows an exemplary drawing for the typical ion traces from the Thomson parabola. The energy and ion species are resolved by the deflection with electric and magnetic fields. The vertical, horizontal, and oblique lines through origin in the drawing are, respectively, corresponding to constant particle energy per charge, constant momentum per charge, and constant velocity. Since the multiple ion species are usually produced by relativistic laser–plasma interactions, the resolving of ion energy and ion species by the Thomson parabola is an essential method.

## Practices

1. What is necessary for vacuum pumps and gauges to construct oil-free vacuum system reaching vacuum level of about $10^{-5}$ Torr?
2. Explain necessary vacuum technologies for reaching vacuum level of about $10^{-8}$ Torr.

3. What is proper working vacuum level for Pirani gauges and ion gauges?
4. Specify a mirror for Ti:Sapphire laser pulse reflecting with p-polarization and having high thermal loading.
5. Design focusing optics to reach $10^{18}$ W $cm^{-2}$ intensity with a laser pulse having 10 mJ energy and 20 fs pulse duration.
6. Specify an OAP mirror to focus a collimated laser pulse having diameter of 10 cm to a focal size of 5 μm.
7. Design a layout of the focal monitoring system to measure the focal spot of the practice #5 and #6.
8. What are the critical densities for 400, 800, and 1600 nm laser pulse in units of electrons $cm^{-3}$?
9. Explain the difference between CCD and CMOS sensors.
10. Why are the back-illuminated cameras more sensitive than the front-illuminated cameras?
11. Explain how EMCCD and ICCD amplify image signals.
12. What would be a proper spectrometer geometry for 50 nm light? Design a spectrometer for the 50 nm light.
13. Calculate Abel inverse transform of a Gaussian profile with 50 μm width in full width at half maximum.
14. Calculate deflection angle when an electron of 1 GeV energy passes through a uniform magnetic field with 1 T magnetic field strength and 15 cm length.
15. Calculate horizontal and vertical deflection angle for a 10 MeV proton passing through a 10 cm long vertical uniform electric field of 10 kV $cm^{-1}$ and magnetic field of 0.5 T.

## References


[1] Leybold 2023 Leybold fundamentals of vacuum technology (https://leybold.com/en/knowledge/vacuum-fundamentals)

[2] Overview of Vacuum (https://pfeiffer-vacuum.com/en/know-how/introduction-to-vacuum-technology/general/overview-of-vacuum/)

[3] Chambers A, Fitch R K and Halliday B S 1998 *Basic Vacuum Technology* (Boca Raton, FL: CRC Press) 2 edn

[4] Hucknall D J 2013 *Vacuum Technology and Applications* (Amsterdam: Elsevier Science)

[5] Useful information on positive displacement pumps (https://michael-smith-engineers.co.uk/resources/useful-info/positive-displacement-pumps)

[6] Vacuum pump: what is it? How does it work? Types of pumps (https://iqsdirectory.com/articles/vacuum-pump.html)

[7] Gaede W 1915 Die diffusion der gase durch quecksilberdampf bei niederen drucken und die diffusionsluftpumpe *Ann. Phys.* **351** 357

[8] How does a turbomolecular pump work (https://leybold.com/en/knowledge/vacuum-fundamentals/vacuum-generation/how-does-a-turbomolecular-pump-work)

[9] O'Hanlon J F 2003 *A User's Guide to Vacuum Technology* (New York: Wiley)

[10] Plücker 1858 Fortgesetzte beobachtungen über die elektrische entladung *Ann. Phys.* **181** 67

[11] 5 tips and tricks for choosing the right vacuum gauges (https://leybold.com/en/knowledge/blog/five-tips-and-tricks-for-choosing-the-right-vacuum-gauges)
[12] 2023 What is the working principle of the Pirani gauge? (https://sens4.com/pirani-working-principle.html)
[13] Bayard R T and Alpert D 1950 Extension of the low pressure range of the ionization gauge *Rev. Sci. Instrum.* **21** 571
[14] *News | How Does an Ion Gauge Work?—Arun Microelectronics* (https://arunmicro.com/news/how-does-an-ion-gauge-work/)
[15] *Ultra-High Vacuum Fittings: What Are Flanges Used For?* (https://vacuumscienceworld.com/blog/what-are-flanges-used-for)
[16] *Optical Coatings* (https://newport.com/n/optical-coatings)
[17] Korea Electro-Optics (https://keoc.kr/?lang=en)
[18] *Corning® ULE® 7973 Low Expansion Glass Optical Materials Product Information Specialty Materials Division* n.d.
[19] 2016 ULE ® Corning Code 7972 Ultra Low Expansion Glass Advanced Optics and Materials ULE ® Corning Code 7972 Ultra Low
[20] *Corning ® HPFS ® 7979, 7980, 8655 Fused Silica Optical Materials Product Information*, (n.d.)
[21] *Interactive Abbe Diagram | SCHOTT* (https://schott.com/en-us/interactive-abbe-diagram)
[22] *Off-Axis Parabolic Mirror Selection Guide | Edmund Optics* (https://edmundoptics.com/knowledge-center/application-notes/optics/off-axis-parabolic-mirror-selection-guide/)
[23] Kevin *An Introduction to Off-Axis Parabolic Mirrors OPTI 521 Tutorial Written by: Kevin Newman* (https://wp.optics.arizona.edu/optomech/wp-content/uploads/sites/53/2016/10/521_Tutorial_Newman_Kevin.pdf)
[24] *What Is a Working Distance (W.D.)? | Learn about Microscope | Olympus* (CRC Press) 2019
[25] Kruer W L 2019 *The Physics of Laser Plasma Interactions* (Boca Raton, FL: CRC Press)
[26] de Laval C G P 1894 Steam-Turbine *US Patent* US522066A
[27] Aniculaesei C, Kim H T, Yoo B J, Oh K H and Nam C H 2018 Novel gas target for laser wakefield accelerators *Rev. Sci. Instrum.* **89** 025110
[28] Prencipe I *et al* 2017 Targets for high repetition rate laser facilities: needs, challenges and perspectives *High Power Laser Sci. Eng.* **5** E17
[29] Nash C P and Musselman R L 1975 A variable path-length infrared gas cell *Appl. Spectrosc.* **29** 527–8
[30] *Pulse Valves—Miniature High Speed High Vacuum Dispense | Parker NA* (https://ph.parker.com/us/en/product-list/pulse-valves-miniature-high-speed-high-vacuum-dispense-valve)
[31] *Targetry—SourceLAB* (https://sourcelab-plasma.com/solutions/targetry/)
[32] Leemans W P, Nagler B, Gonsalves A J, Tóth C, Nakamura K, Geddes C G R, Esarey E, Schroeder C B and Hooker S M 2006 GeV electron beams from a centimetre-scale accelerator *Nat. Phys.* **2** 696
[33] Kim I J *et al* 2012 Relativistic frequency upshift to the extreme ultraviolet regime using self-induced oscillatory flying mirrors *Nat. Commun.* **3** 1
[34] Ehret M, De Luis D, Apiñ Aniz J I, Luis Henares J, Lera R, Antonio Pé Rez-Herná Ndez J, Puyuelo-Valdes P, Volpe L and Gatti G 2023 High-repetition-rate solid tape target delivery system for ultra-intense laser-matter interaction at CLPU *Arxiv* 2302.04769
[35] Kim I J *et al* 2013 Transition of proton energy scaling using an ultrathin target irradiated by linearly polarized femtosecond laser pulses *Phys. Rev. Lett.* **111** 165003

[36] Tomita K *et al* 2017 Time-resolved two-dimensional profiles of electron density and temperature of laser-produced tin plasmas for extreme-ultraviolet lithography light sources *Sci. Rep.* **7** 1

[37] Bang W, Quevedo H J, Dyer G, Rougk J, Kim I, McCormick M, Bernstein A C and Ditmire T 2012 Calibration of the neutron detectors for the cluster fusion experiment on the Texas Petawatt laser *Rev. Sci. Instrum.* **83** 063504

[38] Fedeli L, Formenti A, Cialfi L, Pazzaglia A and Passoni M 2018 Ultra-intense laser interaction with nanostructured near-critical plasmas *Sci. Rep.* **8** 1

[39] *CCD vs CMOS | Teledyne DALSA* (https://teledynedalsa.com/en/learn/knowledge-center/ccd-vs-cmos/)

[40] *A Comparision between Back and Front Illuminated Sensors | Scientist Live* (https://scientistlive.com/content/comparision-between-back-and-front-illuminated-sensors)

[41] *Sony Group Portal—News Release—Sony Develops Back-Illuminated CMOS Image Sensor, Realizing High Picture Quality, Ne* (https://sony.com/en/SonyInfo/News/Press/200806/08-069E/)

[42] *Achieving a True Global Shutter with Large Format, Back-Illuminated CMOS | Teledyne Princeton Instruments* (https://princetoninstruments.com/products/technologies-family/lacera/tech-notes/achieving-a-true-global-shutter-with-large-format-back-illuminated-cmos)

[43] *Learn | ICCD and EmICCD: The Basics in Technology* (https://princetoninstruments.com/learn/camera-fundamentals/iccd-and-emiccd-basics)

[44] *Learn | EMCCD: The Basics in Technology* (https://princetoninstruments.com/learn/camera-fundamentals/emccds-the-basics)

[45] Hecht E 2015 *Optics* 5th edn (Pearson)

[46] Cruddace R G, Gursky H and Meekins J F 1985 Optimization of the rowland circle grating for high-resolution astrophysical spectrometers working at soft x-ray and EUV wavelengths *Appl. Opt.* **24** 2987–95

[47] Osborn K D and Callcott T A 1995 Two new optical designs for soft x-ray spectrometers using variable-line-space gratings *Rev. Sci. Instrum.* **66** 3131

[48] Hough P, McLoughlin C, Kelly T J, Harilal S S, Mosnier J P and Costello J T 2009 Time resolved nomarski interferometery of laser produced plasma plumes *Appl. Surf. Sci.* **255** 5167

[49] Kalal M, Slezak O, Martinkova M and Rhee Y J 2010 Compact design of nomarski interferometer and its applicationin diagnostics of coulomb explosions of deuterium clusters *J. Korean Phys. Soc.* **56** 287

[50] Jeong M, Lee K B, Kim K J, Lee M K and Han J B 2014 Gamma-ray full spectrum analysis for environmental radioactivity by HPGe detector *J. Astron. Space Sci.* **31** 317

[51] Cooper R J, Amman M, Luke P N and Vetter K 2015 A Prototype high purity germanium detector for high resolution gamma-ray spectroscopy at high count rates *Nucl. Instruments Methods Phys. Res. Sect. A Accel. Spectrometers, Detect. Assoc. Equip.* **795** 167

[52] Jeon J H *et al* 2016 Measurement of angularly dependent spectra of betatron gamma-rays from a laser plasma accelerator with quadrant-sectored range filters *Phys. Plasmas* **23** 073105

[53] Koester P *et al* 2021 Bremsstrahlung cannon design for shock ignition relevant regime *Rev. Sci. Instrum.* **92** 013501

[54] Allison J *et al* 2016 Recent developments in GEANT4 *Nucl. Instruments Methods Phys. Res. Sect. A Accel. Spectrometers, Detect. Assoc. Equip.* **835** 186

[55] *The Official FLUKA Site: FLUKA Home* (http://fluka.org/fluka.php)

[56] *MCNP® Website* (https://mcnp.lanl.gov/)

[57] Liang E *et al* 2022 A scintillator attenuation spectrometer for intense gamma-rays *Rev. Sci. Instrum.* **93** 63103

[58] Hojbota C I, Kim H T, Shin J H, Aniculaesei C, Rao B S and Nam C H 2019 Accurate single-shot measurement technique for the spectral distribution of GeV electron beams from a laser wakefield accelerator *AIP Adv.* **9** 85229

# Chapter 4

# Laser particle accelerators

Since ultra-intense lasers have attained the laser intensity required for relativistic laser–plasma interactions, particle accelerations with intense lasers have garnered considerable interest. Due to the breakdown problem of radio-frequency (RF) cavities, conventional RF accelerators have a restricted acceleration gradient. In contrast, laser–plasma particle accelerators do not experience breakdown issues because the plasma is readily ionized medium, and the enormous electric field in plasmas can provide orders of magnitude greater acceleration fields than conventional accelerators. This chapter will introduce the fundamental concept of laser particle accelerations and the proper experimental conditions for particle accelerations with ultra-intense laser pulses.

## 4.1 Relativistic laser–plasma interactions

### 4.1.1 Optical field ionization

When an atom is exposed to an intense laser field, the strong electric field of the laser pulse can ionize the electrons. The laser field can induce ionization via multi-photon ionization or tunneling ionization. The Keldysh parameter [1] distinguishes the ionization regime between multi-photon ionization and tunneling ionization; it is denoted by

$$\gamma_{\text{kel}} = \sqrt{\frac{I_p}{2U_p}} = \frac{\omega_L}{eE_0}\sqrt{2m_e I_p} \tag{4.1}$$

where the $\omega_L$ is angular frequency of the laser field, $e$ is the charge of an electron, $E_0$ is the electric field strength of the laser field, $m_e$ is the mass of an electron, and $I_p$ is the ionization potential. $U_p$ is the ponderomotive energy, the cycle-averaged quiver energy of an electron in an electromagnetic field, given by

doi:10.1088/978-0-7503-2703-9ch4

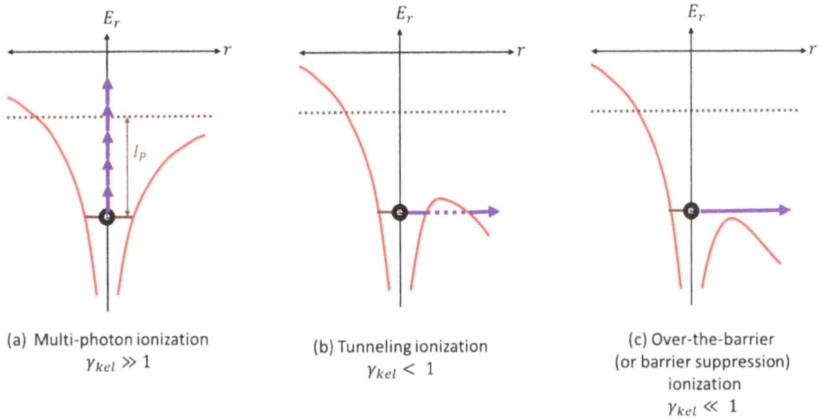


**Figure 4.1.** Three ionization processes: (a) multi-photon ionization, (b) tunneling ionization, and (c) over-the-barrier ionization.

$$U_p = \frac{e^2E_0^2}{4m_e\omega_L^2} \approx 9.33 \times I(10^{14}\quad \mathrm{W\,cm} - 2) \cdot \lambda_L^2(\mu\mathrm{m}^2) \tag{4.2}$$

where the $\lambda_L$ is the wavelength of the laser field. The multi-photon ionization mostly occurs when the ionization potential is much larger than the ponderomotive potential ($\gamma_{\mathrm{kel}} \gg 1$) as shown in figure 4.1. The tunneling ionization dominates as laser intensity and wavelength increase ($\gamma_{\mathrm{kel}} < 1$). The laser field deforms the atomic potential in this condition, and the bound state is near the deformed potential barrier, causing electron tunneling to free space through the shallow potential barrier. When the laser field is extremely strong and greater than the atomic potential, the electron in the bound state is ionized directly by the strong electric field of the laser, a process known as over-the-barrier ionization or barrier-suppression ionization (BSI). Tunneling ionization and over-the-barrier ionization are commonly dominant ionization processes with ultra-intense ultrashort laser pulses having infrared wavelength [2].

The simplest model of ionization is BSI or over-the-barrier ionization that can be considered the classical limit of the tunneling ionization. As the laser field over the critical field, the potential barrier is suppressed and electrons can escape freely. The critical field for the BSI ionization is $F_{\mathrm{critic}} = I_p^2/(4eZ)$, where $Z$ is the ion charge state. The BSI model is only for higher laser field strength than the critical field, which cannot be applied to expect various ionization conditions. When the ponderomotive potential of laser field is comparable to the ionization potential, the process of tunneling ionization becomes the most significant. The tunneling ionization rate can be calculated by solving time-dependent Schrödinger equation, while it can be calculated by Ammosov–Delone–Krainov (ADK) model [3] based on strong field approximation and hydrogen-like atom. The ionization rate in quasi-static regime ($\gamma_{\mathrm{kel}} < 1/2$) according to the ADK model can be written as

$$w_{\mathrm{ADK}}(t) = |\, C_{n*l*} \,|^2 \cdot f(l,\, m) \cdot I_p \left( \frac{2F_i}{F(t)} \right)^{2n*-|m|-1} e^{\left(-\frac{2F_i}{3F}\right)} \tag{4.3}$$

where

$$| C_{n*l*} |^2 = \frac{2^{2n*}}{n^*\Gamma(n^* + l^* + 1)\Gamma(n^* - l^*)} \tag{4.4}$$

$$f(l, m) = \frac{(2l + 1)(1 + | m |)!}{2^{|m|}(| m |)!(l - | m |)!}. \tag{4.5}$$

The $F_i = (2I_p)^{2/3}$ and $F(t)$ are the amplitude of the laser's electric field. Variables $l$ and $m$ are orbital and magnetic quantum number, respectively. The effective principal quantum number is $n^* = Z/(2I_p)^{1/2}$ and effective orbital quantum number is $l^* = n_0^* - 1$, where $n_0^*$ is the effective principal quantum number of the ground state. The total ionization probability is given by

$$P(t) = 1 - e^{-\int_{-\infty}^{t} W_{\mathrm{ADK}}(t^{'})dt^{'}}. \tag{4.6}$$

### 4.1.2 Charged particle motion in a strong laser field

Once atoms are ionized by laser fields, the ionized electrons are driven by the laser field. The motion of an electron in a laser field can be described by Lorentz equation,

$$\frac{d\boldsymbol{p}}{dt} = -e(\boldsymbol{E}_L + \boldsymbol{v} \times \boldsymbol{B}_L) \tag{4.7}$$

where the $\boldsymbol{p} = \gamma m_e \boldsymbol{v}$ is the momentum of an electron, and $\boldsymbol{E}_L$ and $\boldsymbol{B}_L$ are, respectively, the electric field and the magnetic field of a driving laser. $\gamma = 1/\sqrt{1 - \beta^2}$ is the relativistic factor and $\beta = v/c$ is the generalized velocity. For the nonrelativistic case ($\gamma \ll 1,\ v \ll c$), the second term of Lorentz equation is negligible and the electron moves transversally. As the velocity of the electron in the laser field is getting closer to the speed of light, the electron can have longitudinal motion by the $\boldsymbol{v} \times \boldsymbol{B}_L$ term. However, the $\boldsymbol{v} \times \boldsymbol{B}_L$ term only makes displacement of the electron after the laser pulse, so that the Lorentz force with travelling electromagnetic wave cannot transfer the momentum or energy to electrons directly.

The relativistic laser–plasma interaction is defined as the quiver motion of an electron in a strong laser field that is close to the speed of light. The strength of the laser field in relativistic regime can be quantified by the normalized vector potential $a_0$ given by [4]

$$a_0 = \frac{eE_0}{m_e\omega_0 c}. \tag{4.8}$$

The normalized vector potential for 800 nm wavelength is about $2.2 \times 10^{18}$ W cm$^{-2}$, which can be achievable by focusing a TW laser pulse down to about a few microns. The normalized vector potential indicates the ratio between the speed of nonrelativistically oscillating electron in a laser field and the speed of light. Thus, the electrons in a laser field with $a_0 > 1$ move relativistically.

When a nonrelativistic laser pulse is incident to a plasma medium, the electrons in the plasma medium oscillate around the equilibrium position with an angular frequency, called plasma frequency, given by

$$\omega_p = \sqrt{\frac{n_e e^2}{m_e \varepsilon_0}}. \tag{4.9}$$

The plasma interacting with laser field acts as an optical material and the dispersion relation of an electromagnetic wave in a plasma can be written as

$$\omega_0^2 = \omega_p^2 + c^2 k^2 \tag{4.10}$$

where the $k$ is the wave number of the laser field. When the plasma frequency is same as laser frequency, the wave is not travelling ($k = 0$), which corresponds to the electron density of the plasma that is equal to the critical density in equation (3.2). The refractive index of a plasma medium from the dispersion relation can be written as

$$n_p = \sqrt{1 - \frac{\omega_p^2}{\omega_0^2}}. \tag{4.11}$$

The phase velocity and group velocity of the electromagnetic waves in a plasma can be also derived from dispersion relation as

$$\begin{aligned} v_{ph} &= \frac{\omega_0}{k} = \sqrt{c^2 + \frac{\omega_p^2}{k^2}} \\ v_g &= \frac{d\omega_0}{dk} = \frac{c^2}{\sqrt{c^2 + \frac{\omega_p^2}{k^2}}}. \end{aligned} \tag{4.12}$$

The effective mass change of the electron affects the plasma frequency in the relativistic regime [5], as given by

$$\omega_{p\_rel} = \frac{\omega_p}{\sqrt{\gamma}} = \sqrt{\frac{n_e e^2}{\gamma m_e \varepsilon_0}}. \tag{4.13}$$

### 4.1.3 Laser propagation though an under-dense plasma

The refractive index of the plasma medium varies, and a laser pulse propagating through it undergoes spatiotemporal changes. Several well-known phenomena induced by laser modifications in an under-dense plasma medium are introduced in this section.

(1) Self-phase modulation

When an intense laser pulse passes through a gaseous medium, the laser field ionizes the atoms or molecules, resulting in a time-varying refractive index.

The refractive index $n_p = \sqrt{1 - n_e/n_c}$ can be derived from equations (3.2), (4.9), and (4.11). When a neutral gas is ionized by an intense laser field, the electron density changes rapidly, as does the refractive index of the plasma medium. The change in refractive index causes the laser pulse's instantaneous frequency to vary [6]. A laser field's instantaneous frequency in varying refractive index is given by

$$\omega(t) = \frac{d\varphi}{dt} = \frac{d(\omega_0 - kz)}{dt} = \frac{d\left(\omega_0 - \left(\frac{2\pi L n_p}{\lambda_0}\right)\right)}{dt} = \omega_0 - \frac{2\pi L}{\lambda_0}\frac{dn_p}{dt}. \tag{4.14}$$

The varying laser frequency during the ionization of medium is called ionization-induced self-phase modulation (SPM) [7–9]. When the laser intensity increases, the refractive index of the plasma decreases and the laser frequency increases, a phenomenon known as blueshift. As shown in figure 4.2(a), the laser frequency typically increases during the ionization process, and then returns to its original frequency. When

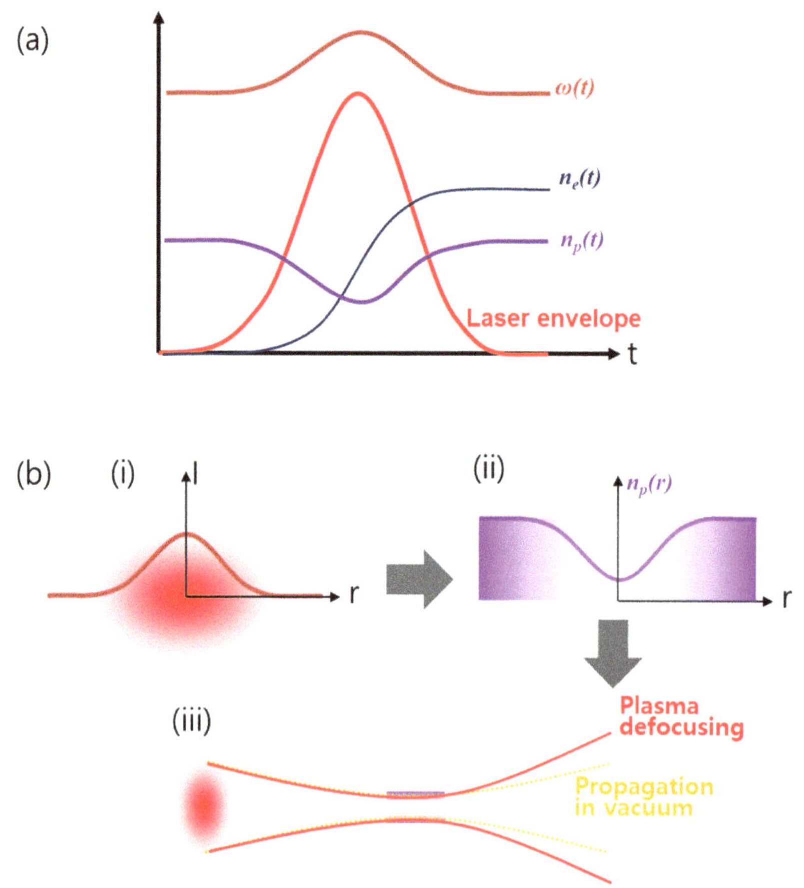


**Figure 4.2.** Principles of (a) SPM induced by ionization process and (b) plasma defocusing.

the laser intensity is greater than the ionization threshold of neutral gases but less than the relativistic regime, this ionization-induced SPM can have a significant effect on laser–plasma interactions, typically in the range of $10^{14}$–$10^{17}$ W cm$^{-2}$ for 800 nm laser wavelength. High-order harmonics and filamentations are the most common in this laser intensity range.

(2) Plasma defocusing

When an intense laser is focused on a gas target, the axial part is more intense than the peripheral part, and the electron density at the center is higher than the electron density off-centered. The spatially central portion of the laser pulse has a lower refractive index than the outer part and the plasma medium acts as a negative lens [10]. As a result, plasma defocusing occurs. The process of the laser pulse being defocused by the higher electron density at the beam center can occur after the ionizing medium. Plasma defocusing takes place in the laser intensity range that causes ionization-induced SPM. The balance between self-focusing by Kerr effect and plasma defocusing can induce filamentation in the air, allowing focused femtosecond laser to propagate over km distances [11, 12].

(3) Relativistic self-focusing

Self-focusing in nonlinear optics is caused by the Kerr effect [13], which occurs because the intense central part has a higher refractive index than the surrounding part. Relativistic self-focusing occurs in a completely distinct physical regime from the Kerr effect when the laser intensity is sufficient to induce relativistic electron motion. At weakly relativistic regime, the plasma medium's index of refraction can be expressed as [5]

$$n_{\text{p\_rel}} = \sqrt{1 - \frac{\omega_{\text{p\_rel}}^2}{\omega_0^2}} = \sqrt{1 - \frac{\omega_{p0}^2}{\omega_0^2 \cdot \gamma}} = \sqrt{1 - \frac{n_e/n_c}{\gamma}} \tag{4.15}$$

where $\omega_{p0}$ is nonrelativistic plasma frequency given by equation (4.9), $n_e$ is electron density profile, $n_c$ is the critical density defined in equation (3.2), and $\gamma$ is the distribution of the relativistic factor. At weakly relativistic regime, the relativistic factor can be approximated as $\gamma = \sqrt{1 + a^2}$, where $a$ is the distribution of the normalize vector potential. For under-dense plasma $n_e \ll n_c$ and weakly relativistic regime a $\ll$ 1, the refractive index in equation (4.15) can be approximated to

$$n_{\text{p\_rel}} \approx 1 - \frac{n_e}{2n_c} \times \frac{1}{\sqrt{1 + a^2}} \approx 1 - \frac{n_e(r, z)}{2n_c} + \frac{n_e \cdot a^2}{4n_c}. \tag{4.16}$$

The third term in equation (4.16) is the term related to the relativistic effect corresponding to the effective mass change. The higher intensity of the laser pulse at the center increases the refractive index by the third term, and the lower electron density at the center caused by the ponderomotive force can also increase the refractive index at the center of the beam by the second term of equation (4.16). As a result, the laser beam can be focused by the higher index of refraction at its center, a phenomenon known as relativistic self-focusing. The parameter denoting the threshold for relativistic self-focusing is commonly referred to as the relativistic critical power, and its mathematical expression is defined as [14]

$$P_{\text{cr}} = \frac{m_e^2 c^5 \omega_0^2}{e^2 \omega_p^2} \approx 17 \left( \frac{\omega_0}{\omega_p} \right)^2 [GW]. \tag{4.17}$$

The relativistic self-focusing plays a crucial role in the laser guidance over distances that significantly exceed the Rayleigh length. This mechanism enables the acceleration of electrons to high energies through laser wakefield acceleration (LWFA).

(4) Photon deceleration and relativistic SPM

The phenomenon of photon deceleration, also known as relativistic SPM [6, 15], can be attributed to the changes in the refractive index caused by relativistic effects. The concept of the relativistic SPM can be comprehended by examining the alterations in the refractive index, as outlined in equation (4.16). The calculation of laser frequency variation can be determined using equation (4.14), which is expressed as

$$\delta\omega(t) = \omega(t) - \omega_0 = -\frac{2\pi L}{\lambda_0} \frac{dn_{p_{\text{rel}}}}{dt} \approx -\frac{2\pi L}{\lambda_0} \frac{n_e \cdot a(t)}{4n_c} \frac{da(t)}{dt}. \tag{4.18}$$

In equation (4.17), it is postulated that the electron density does not vary with time near the laser pulse peak. This assumption is made because the ionization of the medium can be completed at very early phase of the laser pulse. The relativistic SPM exhibits contrasting characteristics compared to the SPM in nonlinear optics. The laser frequency decreases during the rising edge, a phenomenon known as redshift, and increases during the falling edge. Consequently, the alteration of the refractive index due to the relativistic motion of electrons can lead to the occurrence of self-focusing in spatial dimensions and relativistic SPM in temporal dimensions.

## 4.2 Laser electron acceleration

The concept of plasma electron acceleration was initially proposed by J Dawson and T Tajima in 1979 [16]. However, the necessary driving laser or particle beams were not developed until the late 1990s. In the year 2003, a significant achievement was made in the field of laser–plasma electron acceleration [17–19]. This achievement involved the production of mono-energetic focused electrons through the utilization of plasma bubbles, which are also referred to as the blowout regime [20]. Laser–plasma electron acceleration is mostly performed through the utilization of the LWFA mechanism [4, 21, 22]. When a focused laser beam is directed to a gaseous medium at relativistic laser intensity, the electrons within the medium are expelled due to the ponderomotive force exerted by the intense laser. The ponderomotive force is a nonlinear force that a charged particle experiences in an inhomogeneous oscillating electromagnetic field. It causes the particle to move towards the area of the weaker field strength. This occurs because the particle sees a greater magnitude of force during the half of the oscillation period while it is in the area with the stronger field. The net force during its period in the weaker area in the second half of the oscillation does not offset the net force of the first half, and so over a complete cycle this makes the particle move towards the area of lesser force. The ponderomotive force for an electron in a laser field can be defined as the mathematical derivative of the laser intensity [23]:

$$F_p = -\frac{e^2}{4m_e\omega_0^2}\nabla E^2 = -\frac{e^2}{2\,cn_0\varepsilon_0 m_e\omega_0^2}\nabla I. \tag{4.19}$$

The electron driven by a laser pulse with relativistic intensity is expelled away from the peak of the laser pulse by the ponderomotive force. In contrast, the ions remain stationary due to their heavy mass. The expulsion of electrons from their original positions and their subsequent separation from stationary ions results in the creation of a significant electrostatic field. This field exerts a force on the electrons, causing them to move back towards the axis of the laser. The process of separating and restoring electrons occurs repetitively following the propagating laser pulse. This leads to the formation of plasma wave structures with a restoring period that corresponds to the plasma wavelength [24–26]. When the intensity of the laser exceeds a threshold value known as $a_0$, and the duration of the pulse is less than half the plasma wavelength, the plasma wave undergoes deformation and assumes a spherical morphology, commonly referred to as plasma bubbles. The electric field magnitude within the plasma bubble typically reaches 100 gigavolts per meter (GV $m^{-1}$), enabling the acceleration of electrons to energies of 1 giga electron volt (GeV) through LWFA within a distance of approximately 1 cm [27, 28], which is more than 3 orders of magnitude higher than the conventional RF accelerators [29]. Furthermore, the transversal electric field within the plasma bubble serves to maintain the confinement of the electron beam near the axis. This results in the generation of an electron beam that exhibits a pencil-like shape and possesses a small divergence.

The LWFA experiments are performed by focusing an ultra-intense laser pulse into a gas medium. Figure 4.3 depicts the schematic representation of the LWFA experiment. The optimization of electron acceleration in the experiment can be achieved by adjusting the parameters of the laser and medium. In order to achieve a high-energy and high-quality electron beam, it is necessary to optimize both the electron bunch injection process and the acceleration process. The control of the LWFA processes is a highly delicate task due to the fact that the acceleration structure typically is in the dimensions of the plasma wavelength, which is typically on the order of tens of micrometers. The size of the electron bunch is significantly smaller than the plasma wavelength, and the injection process is often highly nonlinear phenomena. Hence, possessing knowledge regarding the bunch injection process and acceleration processes is imperative for conducting the LWFA experiments.

### 4.2.1 Electron bunch injection to plasma wave

The injection of an electron bunch into the plasma wave is an important topic with LWFA because it can determine the quality of the electron beam [30–33].

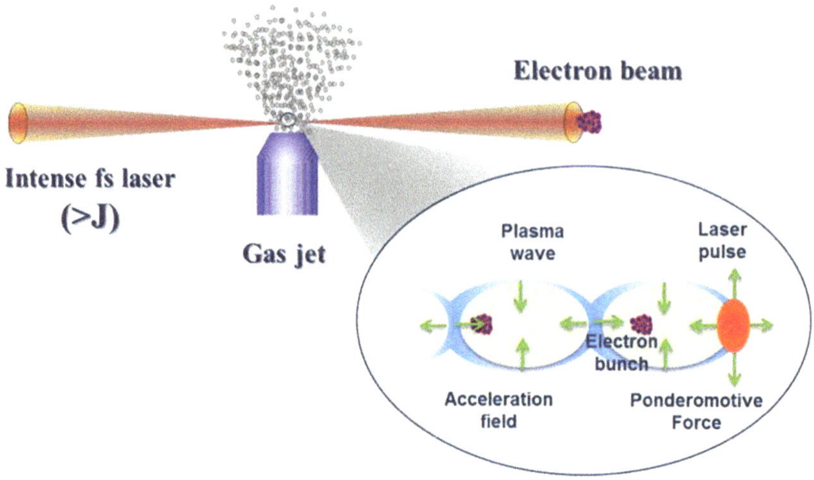


**Figure 4.3.** Schematic drawing of laser wakefield electron acceleration.

The self-injection process is recognized as the most common mechanism for electron injection [34–38]. The phenomenon of self-injection arises when the velocities of electrons surpass the phase velocity of a plasma wave. Electrons that attain sufficient momentum through their interaction with both the ultra-intense laser field and plasma field can achieve longitudinal velocities surpassing the phase velocity of the plasma wave. Consequently, these electrons can be injected and accelerated. The process of self-injection can be comprehended by considering the motion of electrons in phase space within the moving frame corresponding to the phase velocity of the plasma wave. Figure 4.4 depicts the electric field of a plasma wave and presents a phase-space diagram of the electron motion. In the phase-space diagram, an electron with an initial velocity of zero exhibited oscillatory motion with velocities below the injection threshold. The velocity of an electron increases when it interacts with laser and plasma fields, having chances to acquire the necessary velocity ($\Delta v$) for self-injection that can be contained within the separatrix. One potential approach to decrease the injection threshold involves increasing the amplitude of plasma wave or the reduction of the phase velocity of the plasma wave. The presence of elevated plasma densities can create favorable conditions for self-injection due to the resulting higher amplitude of plasma waves and lower phase velocity. Nevertheless, the elevated density may impose restrictions on the energy of the accelerated electron beam, a topic that will be further examined in the following section. Self-injection can be initiated through the implementation of density down ramp or strong self-focusing mechanisms. Both methods effectively decrease the phase velocity of the plasma wave by increasing the plasma wavelength.

The process of self-injection is highly nonlinear and presents challenges in controllability. Numerous schemes have been suggested and demonstrated to enhance the controllability and improve the quality of the injected electron bunches (figure 4.5).

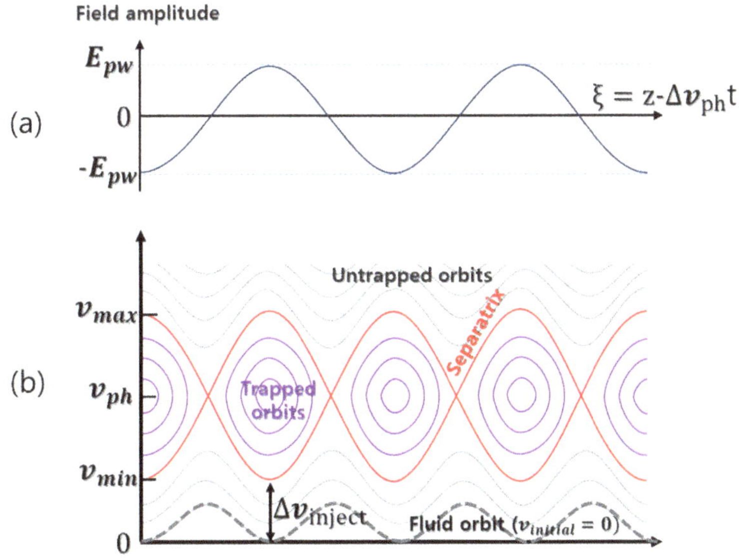


**Figure 4.4.** (a) Field amplitude of plasma wave and (b) phase-space trajectories of electrons in moving frame with the speed of phase velocity of plasma wave.

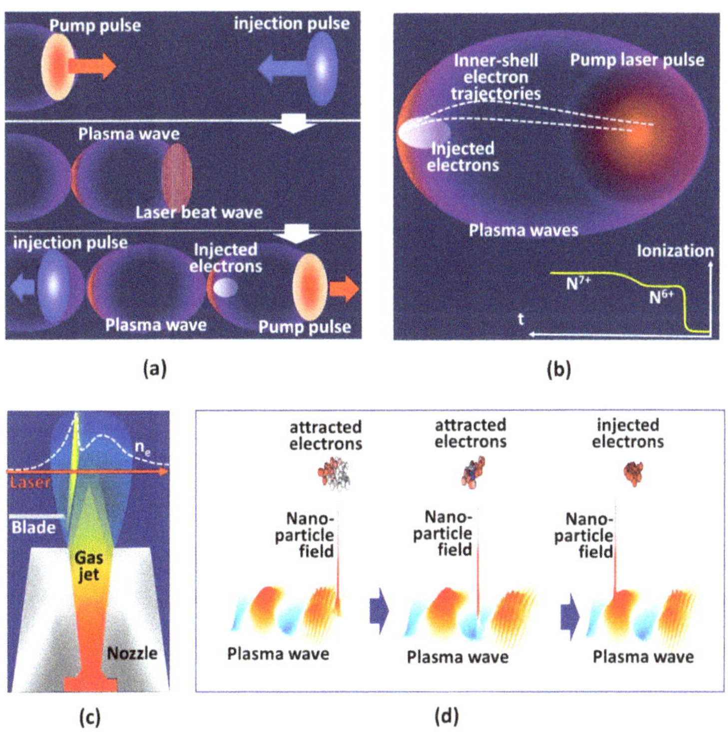


**Figure 4.5.** Schematic drawing for injection methods: (a) colliding pulse injection, (b) ionization injection, (c) density shock injection, and (d) nanoparticle injection (Reproduced from [45]. CC BY 4.0.).

These include colliding pulse injection [39, 40], ionization injection [41–43], density shock injection [31, 42], and nanoparticle insertion method [44–47], among others. The method of colliding laser pulses can enhance the momentum of electrons by generating a beat wave through the interaction between a pump laser pulse and an additional laser pulse for the electron injection. The beat wave has the potential to enhance the electron bunch injection within a confined region by providing additional momentum to electrons within a short window of overlapping the two pulses. The ionization injection techniques involve the utilization of inner-shell ionization of dopant gases, which takes place near the maximum intensity of the laser pulse within the plasma bubble. Most LWFA experiments employ hydrogen or helium gases as the primary plasma medium. The ionization injection scheme utilizes a small fraction of high atomic number ($Z$) gases, such as nitrogen or oxygen, which can facilitate ionization injection through inner-shell ionization processes. The electron trajectories resulting from inner-shell ionization can acquire higher velocities compared to the electrons involved in the formation of the plasma wave, enabling sufficient velocities for injection into the plasma bubble.

The density shock scheme involves the implementation of a short-range density down ramp, achieved by introducing a sharp blade at the early periphery of the supersonic gas jets. This insertion of a sharp blade generates a density shock, which in turn facilitates the injection of the electron bunch through the abrupt expansion of the plasma bubble. The generation of a shock wave by a slender blade can result in the formation of a highly concentrated density peak within a length scale of approximately few tens of microns. This localized density peak has the potential to induce the injection of electron bunches with a high charge in a very confined region. The nanoparticle injection method involves the utilization of particles with dimensions on the order of nanometers for the purpose of injecting electron bunches. The ionization of the nanoparticle occurs by the interaction with the intense laser pulse, leading to the creation of a highly localized nanoparticle field. This field exhibits a strong attraction of vicinal electrons, causing them to travel though the plasma wave. As the nanoparticle reaches the back of the plasma wave, the nanoparticle can liberate the attracted electrons into the plasma wave due to the potential inversion occurring at the back of the plasma wave. The objective of these electron injection schemes is to introduce electron bunches into a confined region in phase space, commonly referred to as localized electron injection. These approaches are intended to minimize the energy spread and emittance of the electron beam. Following the precise injection of electrons in the initial stage of the acceleration medium, it is imperative to maintain effective control over the subsequent acceleration processes to achieve electron beams of high quality and high energy.

### 4.2.2 Acceleration processes

The energy and quality of the accelerated electron beam in LWFA depends on the acceleration processes of LWFA, including laser propagation, dephasing, and pump depletion. This section presents the fundamental formulas necessary for comprehending the LWFA processes in the blowout or bubble regime. For a more comprehensive understanding of the acceleration mechanisms, it is recommended to refer to the provided references [14, 33, 48–54]. Without external guiding, it is necessary for the laser propagation to satisfy the matching condition to achieve stable propagation. It can be achieved by balancing the transverse ponderomotive force with the force exerted by the ion channel. In the case of a nonlinear regime ($a_0 > 2$), the matching conditions can be expressed as follows [14, 49, 50],

$$\begin{aligned} a_0 &\cong 2\left(\frac{P}{P_c}\right)^{1/3} \\ k_pR &\cong k_pw_0 = 2\sqrt{a_0} \\ k_p\tau &\leqslant 2\sqrt{a_0} \end{aligned} \tag{4.20}$$

where the $R$ is the matched blowout radius or matched plasma bubble radius, $k_p$ is the wave number of the plasma wave, $w_0$ is matched focal spot size, and $\tau$ is pulse duration in full width at half maximum (FWHM). The energy gain in LWFA is proportional to the production of the acceleration field strength and acceleration length,

$$\Delta E = eE_zL_{\text{acc}} = \frac{m_ec\,\omega_p\,E_z}{E_0}L_{\text{acc}} \tag{4.21}$$

where $E_0 \;= m_ec\,\omega_p/e$ is the wave-breaking field limit for relativistic cold plasma. For a simple calculation, $E_0$ can be represented as

$$E_0\,(eV/m) \cong 96.2\sqrt{n_0[cm^{-3}]} \tag{4.22}$$

where $n_0$ is unperturbed plasma density. The expression for the averaged acceleration field $E_z$ at matched conditions in three-dimensional nonlinear regime, where spherical plasma waves are generated within the range of $2 \leqslant a_0 \leqslant 4$, can be formulated by

$$\langle E_z\rangle \approx m_ec\,\omega_p\sqrt{a_0}/e. \tag{4.23}$$

The limitation of acceleration length can be determined by either the pump depletion length or the dephasing length. The pump depletion length refers to the distance over which a laser's energy is dissipated to a level insufficient to excite a plasma wave effectively. The depletion of the pump is attributed to pulse etching, also known as pulse erosion, which is associated with the localized loss of laser energy due to the deceleration of photons. The velocity of etching in the nonlinear regime can be expressed as

$$v_{\text{etch}} \approx \frac{c\,\omega_p^2}{\omega_0^2}. \tag{4.24}$$

Thus, the laser will be depleted after propagating the pump depletion length as given by

$$L_{\text{pd}} \approx \frac{c}{v_{\text{etch}}} c\tau \approx \frac{\omega_p^2}{\omega_0^2} c\, \tau. \tag{4.25}$$

The dephasing length refers to the distance over which electrons propagate before experiencing a phase slip, which occurs when the electron bunch reaches the half period of the plasma wave. The phase velocity of a plasma wave in a uniform plasma, when coupled with a stationary laser pulse, is found to be equivalent to the group velocity of the laser pulse, which is lower than the speed of light. The phase velocity of plasma waves can be reduced below the group velocity of the laser pulse when considering the etching of the laser pulse. The expression for the phase velocity of the plasma wave can be formulated as follows:

$$v_\varphi \approx v_g - v_{\text{etch}} \approx c\left(1 - \frac{3\omega_p^2}{2\omega_0^2}\right) \tag{4.26}$$

where $v_g$ is the linear group velocity of laser pulse in a very under-dense uniform plasma ($\omega_p^2 \ll \omega_0^2$). Consequently, the dephasing length can be given by

$$L_{\text{deph}} \approx \frac{c}{c - v_\varphi} R \approx \frac{2\omega_p^2}{3\omega_0^2} R \approx \frac{4\omega_p^2}{3\omega_0^2}\sqrt{a_0}. \tag{4.27}$$

Consequently, when the acceleration electron bunch is limited by the dephasing ($L_{\text{deph}} \geqslant L_{pd}$), the energy gain can be formulated approximately by [14]

$$\Delta E \approx \frac{2}{3} m_e c^2 \left(\frac{\omega_0}{\omega_p}\right)^2 a_0 \approx m_e c^2 \left(\frac{P}{m_e^2 c^5/e^2}\right)^{1/3} \left(\frac{n_c}{n_e}\right)^{2/3} \tag{4.28}$$

where $P$ is the peak power of the driving laser pulse. Equation (4.28) can be expressed in a useful form as follows:

$$\Delta E[GeV] \approx 1.7 \left(\frac{P[TW]}{100}\right)^{1/3} \left(\frac{10^{18}}{n_e[\text{cm}^{-3}]}\right)^{2/3} \left(\frac{0.8}{\lambda_0[\mu\text{m}]}\right)^{4/3}. \tag{4.29}$$

The energy gain increases as plasma density decreases, whereas electron injection and self-guiding demand a necessary plasma density for a given laser power. The required electron density for self-guiding can be calculated as follows:

$$n_e \approx \frac{n_c}{a_0^5}. \tag{4.30}$$

The experimental parameters for a given laser power can be estimated by the consideration of the energy scaling formulas and the self-guiding condition. Table 4.1 presents the estimated parameters of LWFA under different laser power conditions. The laser pulse is configured with a wavelength of 800 nm and an initial pulse duration of 30 fs. The laser pulse's pulse durations are elongated from their initial duration to elongate a pump depletion length with keeping the matching

**Table 4.1.** Estimated parameters of LWFA with various laser power without external guiding.

| | | Driving laser system | | | |
|---|---|---|---|---|---|
| Parameter | UNIT | 10 TW laser | 100 TW laser | 1 PW laser | 10 PW laser |
| Laser energy | J | 0.3 | 3 | 30 | 300 |
| Focused spot size (FWHM) | μm | 10 | 25 | 60 | 130 |
| Laser pulse duration (FWHM) | fs | 30 | 45 | 95 | 200 |
| Plasma density | $10^{18}$ cm$^{-3}$ | 4.5 | 0.7 | 0.14 | 0.03 |
| Laser peak power | TW | 10 | 67 | 316 | 1500 |
| Rayleigh range | μm | 283 | 1770 | 10 198 | 47 873 |
| Peak intensity | $10^{18}$ W cm$^{-2}$ | 10 | 11 | 9 | 9 |
| Normalized vector potential | $a_0$ | 2.2 | 2.3 | 2.0 | 2.1 |
| Critical power | TW | 6.6 | 42.3 | 211.5 | 987.1 |
| Blowout radius | μm | 7.4 | 19.1 | 40.6 | 88.0 |
| Plasma wavelength | μm | 15.7 | 39.9 | 89.2 | 192.8 |
| Dephasing length | mm | 1.9 | 31.6 | 336.8 | 3405.0 |
| Pump depletion length | mm | 3.5 | 33.6 | 354.4 | 3481.4 |
| Achievable energy gain | GeV | 0.3 | 1.9 | 9.3 | 43.4 |

condition. The pulse duration in the chirped pulse amplification laser system can be stretched through the manipulation of laser chirps, which can introduce additional effects on the acceleration process and laser propagation [55, 56]. The laser peak powers were maintained at approximately 1.5 times the critical power for laser guiding, while the normalized vector potentials were kept being approximately 2. Based on the estimation by the analytic formulas, the focusing geometry, medium length, plasma density, and detection range of a dipole magnet electron spectrometer can be designed, while it should be noted that the estimated electron energy may be subject to overestimation. The LWFA technique necessitates a mere centimeter-scale acceleration distance to generate GeV electron beams using laser systems with a power of 100 terawatts (TW). By utilizing a petawatt (PW) laser pulse, it is possible to achieve electron energies at the level of 10 GeV with an acceleration length of only a few tens of centimeters [57]. The LWFA employing a 10 PW laser has the capability to achieve energy levels in the tens of GeV range within the self-guiding regime. However, the acceleration length required for this achievement is on the order of several meters. The establishment of a uniform and stable plasma medium at the meter scale presents a formidable task.

The achievable electron energy can be increased through the implementation of external guiding techniques, such as capillary discharge medium [27, 57, 58] or laser–plasma wave guide [59–62]. This approach serves to alleviate the limitations imposed by the self-guiding condition. Table 4.2 shows the parameters associated with LWFA when implemented with external guiding. In instances of external guidance, the laser peak powers were adjusted to approximately 50% of the critical power,

**Table 4.2.** Estimated parameters of LWFA with various laser power with external guiding.

| | | Driving laser system | | | |
|---|---|---|---|---|---|
| Parameter | Unit | 10 TW laser | 100 TW laser | 1 PW laser | 10 PW laser |
| Laser energy | J | 0.3 | 3 | 30 | 300 |
| Focused spot size (FWHM) | μm | 15 | 30 | 75 | 150 |
| Laser pulse duration (FWHM) | fs | 30 | 60 | 120 | 260 |
| Plasma density | $10^{18}$ cm$^{-3}$ | 1.5 | 0.3 | 0.06 | 0.013 |
| Laser peak power | TW | 10 | 50 | 250 | 1154 |
| Rayleigh range | μm | 637 | 2549 | 15 934 | 63 736 |
| Peak intensity | $10^{18}$ W cm$^{-2}$ | 4.5 | 5.7 | 4.5 | 5.2 |
| Normalized vector potential | $a_0$ | 1.5 | 1.6 | 1.5 | 1.6 |
| Critical power | TW | 19.7 | 98.7 | 493.5 | 2277.9 |
| Blowout radius | μm | 10.5 | 24.7 | 52.3 | 116.5 |
| Plasma wavelength | μm | 27.3 | 61.0 | 136.3 | 292.8 |
| Dephasing length | mm | 8.1 | 95.8 | 1012.7 | 10 406.9 |
| Pump depletion length | mm | 10.4 | 104.4 | 1044.4 | 10 444.1 |
| Achievable energy gain | GeV | 0.6 | 3.0 | 15.1 | 69.5 |

while the normalized vector potentials were kept being around 1.5, indicating the quasi-linear regime. The utilization of external guiding compared to self-guiding results in an increase of achievable electron energy of about 50% for a given initial laser peak power. To enhance the energy of electrons through external guiding, it is necessary to decrease the electron injection threshold by employing injection schemes such as down-ramp injection or ionization injection. Even though external guidance can enhance electron energy, it is necessary to extend the required acceleration length as well. Achieving elongation of capillary discharge medium beyond lengths of tens of centimeters poses significant challenges [63]. With the rapid increase in the peak power of laser systems, it is possible to enhance the achievable electron energy to levels exceeding 10 GeV. Nevertheless, the correlation between laser power and the achievable electron energy is not linear, and the production of a lengthy accelerator medium necessary for this purpose is becoming increasingly challenging. Hence, the utilization of multi-stage LWFA technology [28, 64–70] or phase-locked acceleration [71–75] can be an effective method to enhance electron energies.

## 4.3 Particle-in-cell simulation method for laser particle accelerations

While analytic formulas can offer approximate estimations of relativistic laser–plasma interactions, the intricate plasma dynamics involved in highly nonlinear behaviors within intense laser fields cannot be adequately assessed using the simple formulas. Therefore, it is of great significance to employ self-consistent computer simulations to study the interaction between ultra-intense laser fields and relativistic plasmas. These simulations play a crucial role in experiment design and the interpretation of experimental data. To accurately simulate the interactions between laser and plasma, it is necessary to simultaneously calculate the Maxwell equations for the laser and particle fields, as well as the Lorentz equation for particle motion. This calculation involves considering a total of $6N$ parameters, which account for the three spatial dimensions and three momentum dimensions of $N$ particles within the plasma. This task indeed seems impossible to accomplish due to the limitations in computational power. Hence, reasonable simplifications of the calculations are inevitable.

Particle-in-cell (PIC) calculations [76–81] are widely used in numerical simulations of relativistic laser–plasma interactions, including processes such as LWFA [79, 82], ion acceleration [83–85], and relativistic harmonic generation [86, 87]. This method represents electromagnetic fields on a grid and particles through a sample of macroparticles. For a comprehensive understanding of the PIC method, it is recommended to consult the references in this section. Figure 4.6 and supplementary video clip (PIC_LWFA, available to download from https://doi.org/10.1088/978-0-7503-2703-9) show an example of PIC simulation results for LWFA. The PIC simulation was performed using OSIRIS code [77] in two dimensions. The PIC simulation can calculate the laser pulse evolution, density, momentum, and energy distributions of particles in the plasma. The laser field strength, electron density, and particle energy were presented in the figure and video. In the video clip, the self-injection and acceleration processes can be configured clearly. Consequently, the PIC simulations are an essential numerical method to understand relativistic laser–plasma interactions.

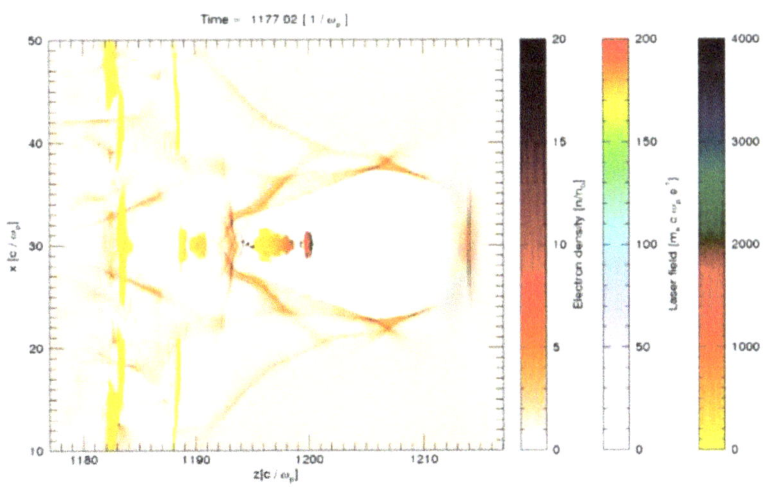


**Figure 4.6.** An example of a PIC simulation result for LWFA. Electron density, laser field strength, and energy of injected electron bunches are overlaid.

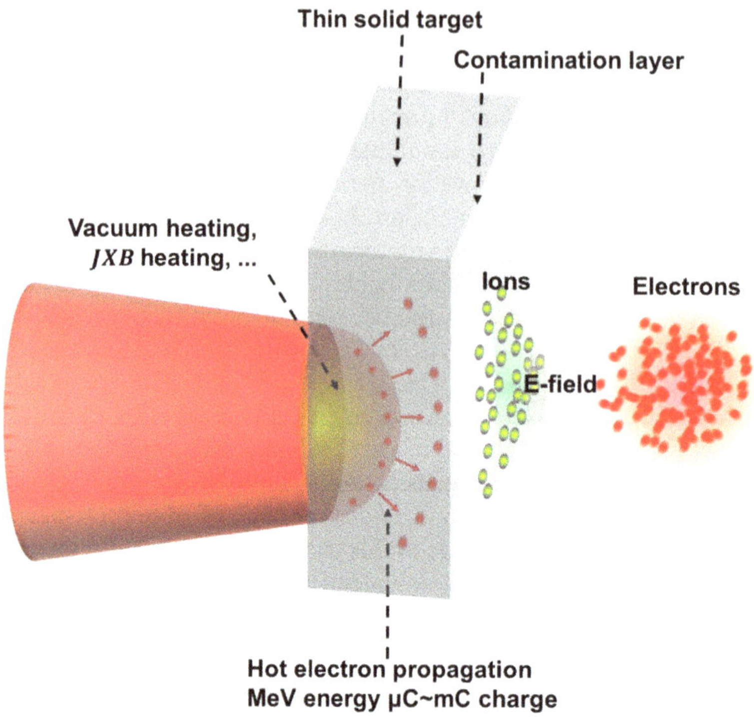


**Figure 4.7.** Schematic drawing of target normal sheath acceleration mechanism.

## 4.4 Laser ion acceleration

The proton and ions are significantly heavier than the electron, which cannot be accelerated to relativistic velocity by the electric field of current ultra-intense lasers and is immobile for the short duration of the intense laser field. The acceleration of ions can occur through the utilization of a strong electric field that is produced by the collective displacement of electrons in an over-dense plasma [83, 84, 88–90]. The ions are accelerated in the plasma until charge neutrality is restored and the ions are emitted following fast electrons. The laser ion acceleration requires an ultra-intense laser pulse with an intensity over $10^{19}$ W cm$^{-2}$ that is focused on a thin over-dense plasma target. Since laser ion acceleration requires higher intensity than LWFA and does not necessitate a lengthy propagation length, short focal length off-axis parabolic mirrors are typically employed as the focusing optics. The emission of protons from thin foil targets is not limited to hydrogen-containing chemical compositions of the foil because protons can also originate from impurities, such as thin layers of water or hydrocarbons, that are commonly found on solid surfaces. The mechanisms of laser ion acceleration vary depending on factors such as laser intensity, pulse duration, target thickness, and plasma density.

The most common method for laser ion acceleration using ultra-intense laser pulses is known as target normal sheath acceleration (TNSA) [88, 90–92]. This phenomenon arises when a thin solid target is irradiated by an ultra-intense laser pulse having an intensity over $10^{19}$ W cm$^{-2}$ ($a_0 \gg 1$). The TNSA mechanism can be described in multiple steps (figure 4.7): formation of energetic electron sheets, electron transport, sheath field formation, and ion acceleration. There are several heating processes induced by the interaction between an ultra-intense laser and an over-dense plasma, while vacuum heating and $J \times B$ heating can be the dominant mechanism for the formation of energetic electrons in the forward direction. The vacuum heating, also known as Brunel heating, occurs at the very steep density gradient of plasma when a p-polarized laser field is obliquely incident on a steplike vacuum–plasma boundary. In a half cycle of a laser period, the electrons move away from the target and into the vacuum, turn around, and are then accelerated back towards the target. When accelerated electrons enter the over-dense plasma target, the laser field diminishes exponentially in the region of over-critical density, and the electrons retain most of their kinetic energy into the plasma. The $J \times B$ heating is brought about by the longitudinal motions of electrons in a relativistic laser intensity due to the $v \times B$ term in the Lorentz force. The electron momentum transfer into plasma is also possible due to the critical density surface as vacuum heating, whereas the electron sheets by $J \times B$ heating occur twice per laser cycle, i.e. $2\omega$ frequency. In addition, resonance absorption near the critical density surface, inverse bremsstrahlung, and collisional heating processes can affect the properties of over-dense plasma and the ion acceleration processes.

The peak current of the energetic electron sheets is on the order of megaamperes (MA), which significantly exceeds the Alfvén limit [93]. The Alfvén limit indicates that the maximum current that can freely propagate in a vacuum is limited by the self-induced azimuthal magnetic field. However, the extremely high current by the relativistic electron sheets in an over-dense plasma plasma can be feasible due to return currents that compensate the self-induced magnetic fields. In addition, a slow drift of cold background electrons can facilitate compensation, given that the background density of an over-dense plasma is excessively higher than the relativistic electron density. Thus, if the target is sufficiently thin, typically a few μm, most relativistic electrons can propagate to the target's rear surface and become trapped due to charge separation, which creates a strong sheath field at the rear surface. The sheath field can be expressed as

$$E_s \approx \frac{k_B T_e}{e\lambda_D} \approx \sqrt{4\pi n_e k_B T_e} \tag{4.31}$$

where $k_B$ is Bolzman constant, $T_e$ is electron temperature, and $\lambda_D$ is Debye length. The Debye length is the characteristic length scale at which mobile charge carriers in plasmas screen out electric fields, indicating the distance over which significant charge separation is possible. The Debye length is given by

$$\lambda_D = \sqrt{\frac{\varepsilon_0 k_B T_e}{e^2 n_e}}. \tag{4.32}$$

For a laser intensity over $10^{19}$ W cm$^{-1}$ and few μm thick target, the sheath field is in order of $10^{12}$ V m$^{-1}$ (= 1 TV m$^{-1}$ = 1 MV μm$^{-1}$), which is sufficient to ionize the atoms in the contamination layer, and the ions can be accelerated according to their charge-to-mass ratio in the sheath field. In the TNSA regime, the achievable ion energy is known to be proportional to the laser intensity to the power of 0.5–1. Since the direction of the electric field in the plasma sheath is perpendicular to the surface of the target, the ions are accelerated in the target normal direction, hence the name 'target normal sheath acceleration.'

The ultra-intense laser pulse reflected by the critical density surface, the light pressure can push the electrons, which results in a strong charge-separation field that can efficiently accelerate the ions, so-called radiation pressure acceleration (RPA) [22, 92, 94–97]. There are two regimes of RPA: light sailing (LS) regime for ultra-thin target and hole-boring regime for thicker target than LS-RPA. Here, LS-RPA mechanism will be described as an example of RPA mechanism. For the LS-RPA ion acceleration, a circularly polarized laser pulse with extremely high laser intensity ($10^{21}$ W $cm^{-2}$) is required to be normally incident to the ultra-thin solid target of thickness of about tens of nanometers to avoid plasma heating processes and transfer sufficient light pressure to the plasma slab. The irradiated region of an ultra-thin foil is first compressed into a plasma slab, and then propelled forward as shown in figure 4.8. If the target thickness is too thin, then the light pressure is diminished due to the relativistic transparency. Therefore, the target should thicker than the threshold of relativistic transparency, of which condition is given as

$$\sigma = \frac{n_e d}{n_c \lambda} \gtrsim \frac{a_0}{\sqrt{2}\pi} \tag{4.33}$$

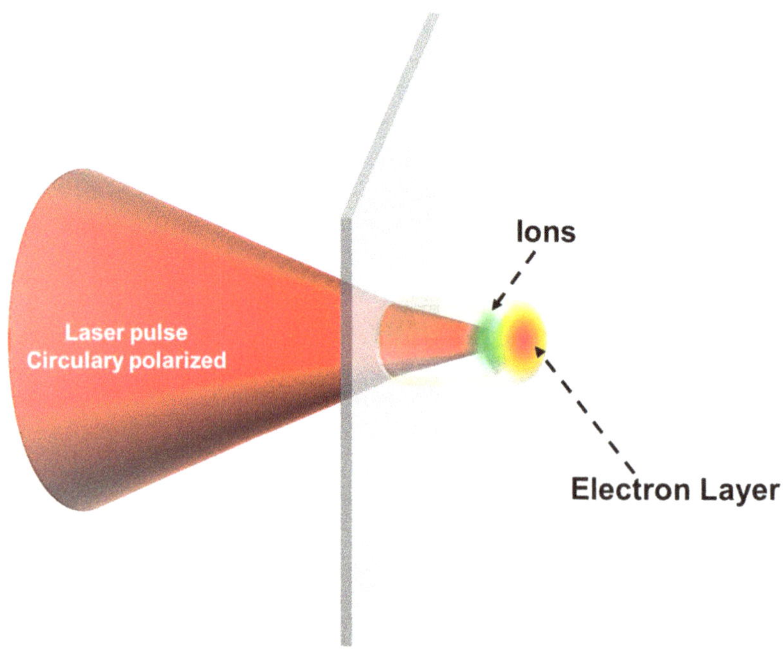


**Figure 4.8.** Schematic drawing of LS-RPA mechanism.

where $\sigma$ indicates the normalized areal density of the target electrons and $d$ is target thickness. If the target thickness is too thick, then the reaching energetic electrons to the rear surface can be disrupted. Thus, the optimum target thickness for LS-RPA can be determined by balancing the target areal density and laser intensity.

When the electrostatic field between the propelled electrons and ions balances with the light pressure of the driving laser pulse, the ions can be accelerated continuously and stably in the presence of ultra-intense laser field. The achievable ion energy per nucleon in LS-RPA regime can be expressed as

$$E_{\mathrm{LS}} = m_i c^2 \frac{F^2}{2(F+1)} \tag{4.34}$$

where the normalized laser fluence $F$ is given by

$$F \equiv \frac{2}{\sigma c^2} \int_0^{\infty} I(t')dt'. \tag{4.35}$$

The normalized laser fluence can be approximated for easy calculations as

$$F \approx \frac{2.2F_L[10^8 \text{ J cm}^{-2}]}{\rho_i\left[\text{g/cm}^{-3}\right] \cdot d_t\left[10 \text{ nm}\right]}. \tag{4.36}$$

Even though RPA requires demanding laser and target condition, the RPA is attracting lots of research efforts for future development of laser ion acceleration because the ion energy can be linearly scaled with the laser intensity, as well as possibility to generate mono-energetic ion beam.

## Practices

1. Calculate the Keldysh parameter for an 800 nm laser pulse with a peak intensity of $1 \times 10^{14}$ W cm$^{-2}$ ionizing a hydrogen atom.
2. What is required laser intensity corresponding to critical field strength for BSI of the second electron of helium atom?
3. Calculate ionization probability of a hydrogen atom when an 800 nm laser pulse with pulse duration of 30 fs is focused on the atom with a peak intensity of $1 \times 10^{14}$ W cm$^{-2}$.
4. What is normalized vector potential $a_0$ value for an 800 nm laser pulse with a laser intensity of $1 \times 10^{19}$ W cm$^{-2}$?
5. What is a laser intensity at 800 nm wavelengths for $a_0$ value of 100?
6. Calculate plasma wavelength for an electron density of $1 \times 10^{19}$ cm$^{-3}$.
7. What is critical density for a laser pulse with a wavelength of 400 nm?
8. Explain the difference between SPM in nonlinear optics regime and relativistic regime.
9. Calculate critical power for relativistic self-focus when an 800 nm laser pulse propagates though a plasma with electron density of $1 \times 10^{18}$ cm$^{-3}$.
10. When a laser pulse, having 1 J energy, 45 fs pulse duration (FWHM), and 800 nm wavelength, is focused into a plasma having electron density of

$1 \times 10^{18}$ cm$^{-3}$ with a focused spot size of 20 μm (FWHM), calculate following values:
   (a) the plasma wavelength
   (b) the laser intensity and normalized vector potential
   (c) critical power for self-focusing
   (d) Blowout radius
   (e) Pump depletion length
   (f) Dephasing length
   (g) Achievable energy
11. Calculate the Debye length and sheath field strength for an electron temperature of 1 MeV and electron density of $1.5 \times 10^{21}$ cm$^{-3}$.
12. Calculate normalized areal density for a target having electron density of 10 times critical density and thickness of 20 nm irradiated by 800 nm laser pulse.
13. Calculate required minimum target thickness to maintain radiation pressure for an 800 nm laser pulse of $1 \times 10^{21}$ W cm$^{-2}$ intensity focused on a plasma with an electron density of $1 \times 10^{22}$ cm$^{-3}$.
14. Calculate normalized laser fluence and achievable proton energy in LS-RPA regime when an 800 nm 20 fs laser pulse with a focused intensity of $1 \times 10^{22}$ W cm$^{-2}$ irradiates an ultra-thin solid target with $n_e = 10\, n_c$ and thickness of 10 nm.

## References

[1] Keldysh L V 1965 Ionization in the field of a strong electromagnetic wave *J. Exp. Theor. Phys.* **20** 1945

[2] Delone N B and Krainov V P 1998 Tunneling and barrier-suppression ionization of atoms and ions in a laser radiation field *Phys.-Usp.* **41** 469

[3] Ammosov M V, Delone N and Krainov V P 1986 Tunnel ionization of complex atoms and of atomic ions in an alternating electromagnetic field *Zh. Ekdp. Teor. Fiz.* **91** 2008–13

[4] Gibbon P 2005 *Short Pulse Laser Interactions with Matter: an Introduction* (London: Imperial College Press)

[5] Cros B 2014 Laser-driven plasma wakefield: propagation effects *Cas-CERN Accel. Sch. Plasma Wake Accel. 2014, Proc.* **001** 207

[6] Giulietti A and Giulietti D 2015 Self-phase modulation in various regimes of intense laser-plasma interactions *J. Plasma Phys.* **81** 495810608

[7] Yablonovitch E 1974 Self-phase modulation and short-pulse generation from laser-breakdown plasmas *Phys. Rev.* A **10** 1888–95

[8] Kim H T, Kim I J, Hong K H, Lee D G, Kim J H and Nam C H 2004 Chirp analysis of high-order harmonics from atoms driven by intense femtosecond laser pulses *J. Phys. B: At. Mol. Opt. Phys.* **37** 1141

[9] Kim H T, Lee D G, Hong K H, Kim J H, Choi I W and Nam C H 2003 Continuously tunable high-order harmonics from atoms in an intense femtosecond laser field *Phys. Rev.* A **67** 051801

[10] Kim H T, Kim I J, Lee D G, Hong K-H, Lee Y S, Tosa V and Nam C H 2004 Optimization of high-order harmonic brightness in the space and time domains *Phys. Rev. A—At. Mol. Opt. Phys.* **69**
[11] Durand M *et al* 2013 Kilometer range filamentation *Opt. Express* **21** 26836
[12] Thul D *et al* 2021 The mobile ultrafast high energy laser facility—a new facility for high-intensity atmospheric laser propagation studies *Opt. Lasers Eng.* **140** 106519
[13] Boyd R W 2020 *Nonlinear Optics* 4th edn (New York: Academic)
[14] Lu W, Tzoufras M, Joshi C, Tsung F, Mori W, Vieira J, Fonseca R and Silva L 2007 Generating multi-GeV Electron bunches using single stage laser wakefield acceleration in a 3D nonlinear regime *Phys. Rev. Spec. Top.—Accel. Beams* **10** 061301
[15] Bingham R, Mendonça J T and Dawson J M 1997 Photon landau damping *Phys. Rev. Lett.* **78** 247
[16] Tajima T and Dawson J M 1979 Laser electron accelerator *Phys. Rev. Lett.* **43** 267
[17] Faure J, Glinec Y, Pukhov A, Klselev S, Gordienko S, Lefebvre E, Rousseau J P, Burgy F and Malka V 2004 A laser–plasma accelerator producing monoenergetic electron beams *Nature* **431** 541
[18] Geddes C G R, Toth C, Van Tilborg J, Esarey E, Schroeder C B, Bruhwiler D, Nieter C, Cary J and Leemans W P 2004 High-quality electron beams from a laser wakefield accelerator using plasma-channel guiding *Nature* **431** 538
[19] Mangles S P D *et al* 2004 Monoenergetic beams of relativistic electrons from intense laser–plasma interactions *Nature* **431** 535
[20] Lu W, Huang C, Zhou M, Mori W B and Katsouleas T 2006 Nonlinear theory for relativistic plasma wakefields in the blowout regime *Phys. Rev. Lett.* **96** 165002
[21] Mourou G A, Tajima T and Bulanov S V 2006 Optics in the relativistic regime *Rev. Mod. Phys.* **78** 309
[22] Tanaka K A *et al* 2020 Current status and highlights of the ELI-NP research program *Matter Radiat. Extremes* **5** 024402
[23] Bituk D R and Fedorov M V 1999 Relativistic ponderomotive forces *J. Exp. Theor. Phys.* **89** 640
[24] Tonks L and Langmuir I 1929 Oscillations in ionized gases *Phys. Rev.* **33** 195
[25] Clayton C E, Joshi C, Darrow C and Umstadter D 1985 Relativistic plasma-wave excitation by collinear optical mixing *Phys. Rev. Lett.* **54** 2343
[26] Esarey E and Pilloff M 1995 Trapping and acceleration in nonlinear plasma waves *Phys. Plasmas* **2** 1432
[27] Leemans W P, Nagler B, Gonsalves a J, Tóth C, Nakamura K, Geddes C G R, Esarey E, Schroeder C B and Hooker S M 2006 GeV electron beams from a centimetre-scale accelerator *Nat. Phys.* **2** 696
[28] Kim H T, Pae K H, Cha H J, Kim I J, Yu T J, Sung J H, Lee S K, Jeong T M and Lee J 2013 Enhancement of electron energy to the multi-GeV regime by a dual-stage laser-wakefield accelerator pumped by petawatt laser pulses *Phys. Rev. Lett.* **111** 165002
[29] Wangler T P 2008 *RF Linear Accelerators* 2nd edn (New York: Wiley)
[30] Faure J, Rechatin C, Norlin A, Lifschitz A, Glinec Y and Malka V 2006 Controlled injection and acceleration of electrons in plasma wakefields by colliding laser pulses *Nature* **444** 737
[31] Buck A *et al* 2013 Shock-front injector for high-quality laser-plasma acceleration *Phys. Rev. Lett.* **110** 185006
[32] Faure J 2014 Plasma injection schemes for laser-plasma accelerators *CAS-CERN Accel. Sch. Plasma Wake Accel. 2014, Proc.* **001** 143

[33] Malka V 2014 Plasma wake accelerators: introduction and historical overview *CAS-CERN Accel. Sch. Plasma Wake Accel. 2014, Proc.* 1–28
[34] Bulanov S V, Pegoraro F, Pukhov A M and Sakharov A S 1997 Transverse-wake wave breaking *Phys. Rev. Lett.* **78** 4205
[35] Esarey E, Hubbard R F, Leemans W P, Ting A and Sprangle P 1997 Electron injection into plasma wakefields by colliding laser pulses *Phys. Rev. Lett.* **79** 2682
[36] Bulanov S, Naumova N, Pegoraro F and Sakai J 1998 Particle injection into the wave acceleration phase due to nonlinear wake wave breaking *Phys. Rev.* E **58** R5257
[37] Kostyukov I, Nerush E, Pukhov A and Seredov V 2009 Electron self-injection in multi-dimensional relativistic-plasma wake fields *Phys. Rev. Lett.* **103** 175003
[38] Corde S, Thaury C, Lifschitz A, Lambert G, Ta Phuoc K, Davoine X, Lehe R, Douillet D, Rousse A and Malka V 2013 Observation of longitudinal and transverse self-injections in laser-plasma accelerators *Nat. Commun.* **4** 1
[39] Faure J, Rechatin C, Norlin A, Lifschitz A, Glinec Y and Malka V 2006 Controlled injection and acceleration of electrons in plasma wakefields by colliding laser pulses *Nature* **444** 737
[40] Kotaki H *et al* 2009 Electron optical injection with head-on and countercrossing colliding laser pulses *Phys. Rev. Lett.* **103** 194803
[41] Pak A, Marsh K A, Martins S F, Lu W, Mori W B and Joshi C 2010 Injection and trapping of tunnel-ionized electrons into laser-produced wakes *Phys. Rev. Lett.* **104** 025003
[42] Thaury C *et al* 2015 Shock assisted ionization injection in laser-plasma accelerators *Sci. Rep.* **5** 16310
[43] Mirzaie M *et al* 2015 Demonstration of self-truncated ionization injection for GeV electron beams *Sci. Rep.* **5** 14659
[44] Shen B, Li Y, Nemeth K, Shang H, Chae Y C, Soliday R, Crowell R, Frank E, Gropp W and Cary J 2007 Electron injection by a nanowire in the bubble regime *Phys. Plasmas* **14** 053115
[45] Cho M H, Pathak V B, Kim H T and Nam C H 2018 Controlled electron injection facilitated by nanoparticles for laser wakefield acceleration *Sci. Rep.* **8** 16924
[46] Aniculaesei C *et al* 2019 Proof-of-principle experiment for nanoparticle-assisted laser wakefield electron acceleration *Phys. Rev. Appl.* **12** 044041
[47] Xu J *et al* 2022 Nanoparticle-insertion scheme to decouple electron injection from laser evolution in laser wakefield acceleration *Sci. Rep.* **12** 1
[48] Tajima T, Yan X Q and Ebisuzaki T 2020 Wakefield acceleration *Rev. Mod. Plasma Phys.* **4** 7
[49] Tsung F S, Lu W, Tzoufras M, Mori W B, Joshi C, Vieira J, Silva L O and Fonseca R A 2006 Simulation of monoenergetic electron generation via laser wakefield accelerators for 5–25TW lasers *Phys. Plasmas* **13** 056708
[50] Lu W, Tzoufras M, Tsung F S, Joshi C, Mori W B, Vieira J, Fonseca R A and Silva L O 2007 Designing LWFA in the blowout regime 2007 IEEE Particle Accelerator Conf. (PAC) pp 3050–1
[51] Esarey E, Schroeder C and Leemans W 2009 Physics of laser-driven plasma-based electron accelerators *Rev. Mod. Phys.* **81** 1229
[52] Malka V and Mora P 2009 Principles of laser–plasma accelerators *C. R. Physique* **10** 106–15
[53] Joshi C 2017 Laser-driven plasma accelerators operating in the self-guided, blowout regime *IEEE Trans. Plasma Sci.* **45** 3134

[54] Kim H T, Pathak V B, Hojbota C I, Mirzaie M, Pae K H, Kim C M, Yoon J W, Sung J H and Lee S K 2021 Multi-GeV laser wakefield electron acceleration with PW lasers *Appl. Sci.* **11** 5831
[55] Kim H T *et al* 2017 Stable multi-GeV electron accelerator driven by waveform-controlled PW laser pulses *Sci. Rep.* **7** 1
[56] Shin J *et al* 2018 Quasi-monoenergetic multi-GeV Electron acceleration by optimizing the spatial and spectral phases of PW laser pulses *Plasma Phys. Control. Fusion* **60** 064007
[57] Gonsalves A J *et al* 2019 Petawatt laser guiding and electron beam acceleration to 8 GeV in a laser-heated capillary discharge waveguide *Phys. Rev. Lett.* **122** 084801
[58] Leemans W P *et al* 2014 Multi-GeV electron beams from capillary-discharge-guided subpetawatt laser pulses in the self-trapping regime *Phys. Rev. Lett.* **113** 245002
[59] Depresseux A *et al* 2015 Table-top femtosecond soft x-ray laser by collisional ionization gating *Nat. Photon.* **9** 817
[60] Lemos N, Cardoso L, Geada J, Figueira G, Albert F and Dias J M 2018 Guiding of laser pulses in plasma waveguides created by linearly-polarized femtosecond laser pulses *Sci. Rep.* **8** 1
[61] Picksley A *et al* 2020 Meter-scale conditioned hydrodynamic optical-field-ionized plasma channels *Phys. Rev.* E **102** 053201
[62] Oubrerie K *et al* 2022 Controlled acceleration of GeV electron beams in an all-optical plasma waveguide *Light: Sci. Appl.* **11** 1
[63] Miao B, Feder L, Shrock J E, Goffin A and Milchberg H M 2020 Optical guiding in meter-scale plasma waveguides *Phys. Rev. Lett.* **125** 074801
[64] Pathak V B, Kim H T, Vieira J, Silva L O and Nam C H 2018 All optical dual stage laser wakefield acceleration driven by two-color laser pulses *Sci. Rep.* **8** 11772
[65] Malka V, Lifschitz a, Faure J and Glinec Y 2006 Staged concept of laser-plasma acceleration toward multi-GeV electron beams *Phys. Rev. Spec. Top.—Accel. Beams* **9** 091301
[66] Kaganovich D, Ting a, Gordon D F, Hubbard R F, Jones T G, Zigler a and Sprangle P 2005 First demonstration of a staged all-optical laser wakefield acceleration *Phys. Plasmas* **12** 100702
[67] Liu J S *et al* 2011 All-optical cascaded laser wakefield accelerator using ionization-induced injection *Phys. Rev. Lett.* **107** 035001
[68] Pollock B B *et al* 2011 Demonstration of a narrow energy spread, ~0.5 GeV electron beam from a two-stage laser wakefield accelerator *Phys. Rev. Lett.* **107** 045001
[69] Steinke S *et al* 2016 Multistage coupling of independent laser-plasma accelerators *Nature* **530** 190
[70] Leemans W and Esarey E 2009 Laser-driven plasma-wave electron accelerators additional resources for physics today *Cit. Phys. Today* **62** 44
[71] Guillaume E *et al* 2015 Electron rephasing in a laser-wakefield accelerator *Phys. Rev. Lett.* **115** 155002
[72] Froula D H *et al* 2018 Spatiotemporal control of laser intensity *Nat. Photon.* **12** 262
[73] Aniculaesei C *et al* 2019 Electron energy increase in a laser wakefield accelerator using up-ramp plasma density profiles *Sci. Rep.* **9** 1
[74] Caizergues C, Smartsev S, Malka V and Thaury C 2020 Phase-locked laser-wakefield electron acceleration *Nat. Photon.* **14** 475
[75] Palastro J P, Shaw J L, Franke P, Ramsey D, Simpson T T and Froula D H 2020 Dephasingless laser wakefield acceleration *Phys. Rev. Lett.* **124** 134802

[76] Pukhov A 2002 Strong field interaction of laser radiation *Rep. Prog. Phys.* **66** 47

[77] Fonseca R A *et al* 2002 OSIRIS: a three-dimensional, fully relativistic particle in cell code for modeling plasma based accelerators *Int. Conf. on Computational Science*

[78] Pritchett P L 2003 Particle-in-cell simulation of plasmas—a tutorial *Space Plasma Simulation* (Heidelberg: Springer) 1–24

[79] Pukhov A 2014 Particle-in-cell codes for plasma-based particle acceleration *CAS-CERN Accel. Sch. Plasma Wake Accel. 2014, Proc.* **001** 181

[80] Vay J L, Lehe R, Vincenti H, Godfrey B B, Haber I and Lee P 2016 recent advances in high-performance modeling of plasma-based acceleration using the Full PIC method *Nucl. Instrum. Methods Phys. Res. Sect. A Accel. Spectrometers, Detect. Assoc. Equip.* **829** 353

[81] Derouillat J *et al* 2018 Smilei: a collaborative, open-source, multi-purpose particle-in-cell code for plasma simulation *Comput. Phys. Commun.* **222** 351

[82] Mori W B 2002 Recent advances and some results in plasma-based accelerator modeling *AIP Conf. Proc.* 647 11

[83] Pukhov A 2001 Three-dimensional simulations of ion acceleration from a foil irradiated by a short-pulse laser *Phys. Rev. Lett.* **86** 3562

[84] Esirkepov T, Yamagiwa M and Tajima T 2006 Laser ion-acceleration scaling laws seen in multiparametric particle-in-cell simulations *Phys. Rev. Lett.* **96** 105001

[85] Djordjević B Z, Kemp A J, Kim J, Simpson R A, Wilks S C, Ma T and Mariscal D A 2021 Modeling laser-driven ion acceleration with deep learning *Phys. Plasmas* **28** 43105

[86] Geissler M, Rykovanov S, Schreiber J, Meyer-Ter-Vehn J and Tsakiris G D 2007 3D simulations of surface harmonic generation with few-cycle laser pulses *New J. Phys.* **9** 218

[87] Teubner U and Gibbon P 2009 High-order harmonics from laser-irradiated plasma surfaces *Rev. Mod. Phys.* **81** 445

[88] Borghesi M 2014 Laser-driven ion acceleration: state of the art and emerging mechanisms *Nucl. Instrum. Methods Phys. Res. Sect. A Accel. Spectrometers, Detect. Assoc. Equip.* **740** 6

[89] Schreiber J *et al* 2006 Analytical model for ion acceleration by high-intensity laser pulses *Phys. Rev. Lett.* **97** 045005

[90] Roth M and Schollmeier M 2016 Ion acceleration—target normal sheath acceleration (CERN Yellow Reports) *Proc. CAS-CERN Accel. Sch. Plasma Wake Accel. Geneva, Switz.* **001** 231

[91] Passoni M, Bertagna L and Zani A 2010 Target normal sheath acceleration: theory, comparison with experiments and future perspectives *New J. Phys.* **12** 045012

[92] Kim I J *et al* 2013 Transition of proton energy scaling using an ultrathin target irradiated by linearly polarized femtosecond laser pulses *Phys. Rev. Lett.* **111** 165003

[93] Alfvén H 1942 Existence of electromagnetic-hydrodynamic waves *Nature.* **150** 405

[94] Kim I J *et al* 2016 Radiation pressure acceleration of protons to 93 MeV with circularly polarized petawatt laser pulses *Phys. Plasmas* **23** 070701

[95] Macchi A, Borghesi M and Passoni M 2013 Ion acceleration by superintense laser-plasma interaction *Rev. Mod. Phys.* **85** 751

[96] Henig A *et al* 2009 Radiation-pressure acceleration of ion beams driven by circularly polarized laser pulses *Phys. Rev. Lett.* **103** 3

[97] Shin S Y, Park S Y and Hahn S J 2019 Investigation of light-sail and hole-boring radiation pressure accelerations upon the interaction of ultra-intense laser pulses with thin targets *J. Korean Phys. Soc.* **75** 968

# Chapter 5

# Radiation sources from laser–plasma interactions

Laser–plasma interactions produce a variety of radiation sources, ranging from radio waves to gamma rays that encompass nearly the entire electromagnetic spectrum. Due to their unique characteristics of short pulse duration, high brightness, compact equipment size, and small source size, laser–plasma radiation sources with short wavelength radiations have received the most attention. This chapter provides an overview of various methods employed in generating radiation sources with short pulse durations and short wavelengths in the extreme-ultraviolet and x-ray ranges, including high-order harmonic generation (HHG), x-ray lasers, and betatron radiation, and other related approaches.

## 5.1 High-order harmonics

HHG is an extremely nonlinear process that generates a large number of harmonics [1, 2]. Depending on the ionization potential of the target medium and the laser intensity, the number of harmonics produced can range from tens to several hundred [3–7]. Typically, the HHG in the non-relativistic regime is generated by coherent interactions between atoms of noble gases and high-intensity laser pulses [8]. In the non-perturbative regime where the laser field can propel electrons beyond the atomic potential barrier, HHG occurs. As depicted in figure 5.1, HHG can be understood in three steps [9]: (i) tunneling ionization of atoms by a strong laser field, (ii) acceleration of electron toward ionic core as laser field direction alternates, and (iii) recombination of the ionized electron with the parent ion and release of acquired energy in the laser field as short-wavelength radiation. The short-wavelength bursts consist of three steps that occur repeatedly every half of the laser period, resulting in the formation of high-order harmonics in spectral domains. Due to the dipole selection rules and the symmetry of the atomic potential, only odd-order harmonics of the driving field can be produced.

doi:10.1088/978-0-7503-2703-9ch5

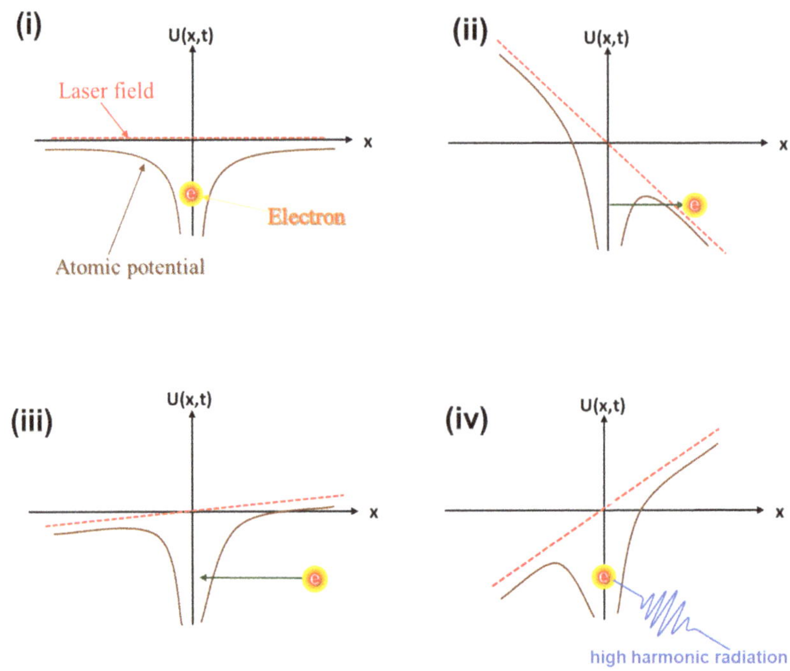


**Figure 5.1.** Schematic drawing for three-step model of HHG.

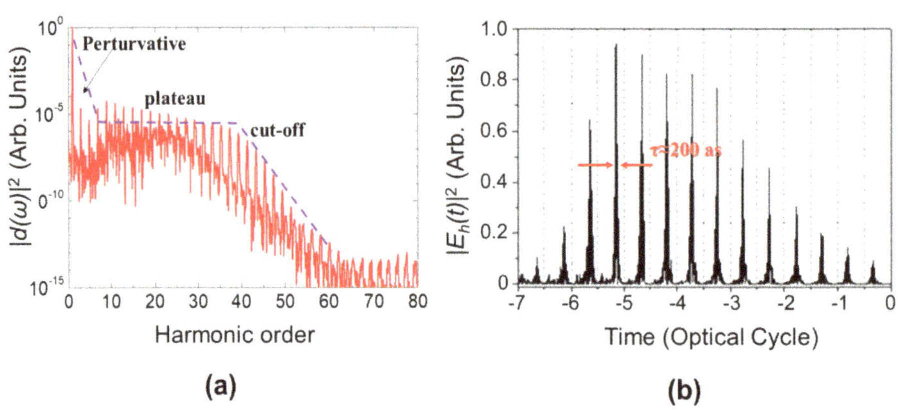


**Figure 5.2.** (a) High-order harmonic spectrum calculated by time-dependent Schrödinger equation with neon atom and a 100 fs laser ($I_L \approx 2 \times 10^{14}$ W cm$^{-2}$) and (b) temporal profiles of high-harmonic radiation calculated by macroscopic propagation HHG calculation with $I_L \approx 2 \times 10^{15}$ W cm$^{-2}$, laser energy = 5 mJ, $\tau$ = −42 fs, target placed at 18 mm before the focal plane, neon medium with 40 Torr density, target length = 9 mm (Reproduced from [14]. ).

The structure of high-order harmonic spectra is shown in figure 5.2(a). The conversion efficiency drops dramatically in the perturbative region where the atomic potential dominates. The harmonic intensities then remain nearly constant with the harmonic order, resulting in a plateau region caused by the non-perturbative interaction in the three-step process. This plateau region is terminated by an abrupt cutoff, and a classical model based on three-step processes [9] was proposed to predict the cutoff energy, which shows the cutoff energy of HHG as given by

$$E_{\mathrm{CUT}} = I_P + 3.17U_P \implies N_{\mathrm{CUT}} = (I_P + 3.17U_P)/(\hbar\omega_L) \tag{5.1}$$

where $E_{\mathrm{CUT}}$ is photon energy of cutoff harmonic order, $p$ is ionization potential, $U_P$ is ponderomotive potential (equation 4.2), $N_{\mathrm{CUT}}$ is cutoff harmonic order, $\hbar$ is reduced Planck constant, and $\omega_L$ is angular frequency of the driving laser pulse. For example, when neon atoms are irradiated with a laser pulse with $I_L \approx 2 \times 10^{14}$ W cm$^{-2}$, the cutoff energy is approximately 59 eV and the cutoff harmonic order is about 39. The depletion of ground-state atoms can also limit the achievable harmonic order. The saturation intensity is defined as the driving laser intensity that is excessively high enough to ionize all of the atoms in the laser field [5, 10], and the maximum harmonic order can be defined as $I_P + 3.17U_{\mathrm{sat}}$, where $U_{\mathrm{sat}}$ is the saturated ponderomotive potential according to the saturation intensity.

Figure 5.3(a) and supplementary animation 'HHG(chap5).wmv' (available to download from https://doi.org/10.1088/978-0-7503-2703-9) depict an experiment

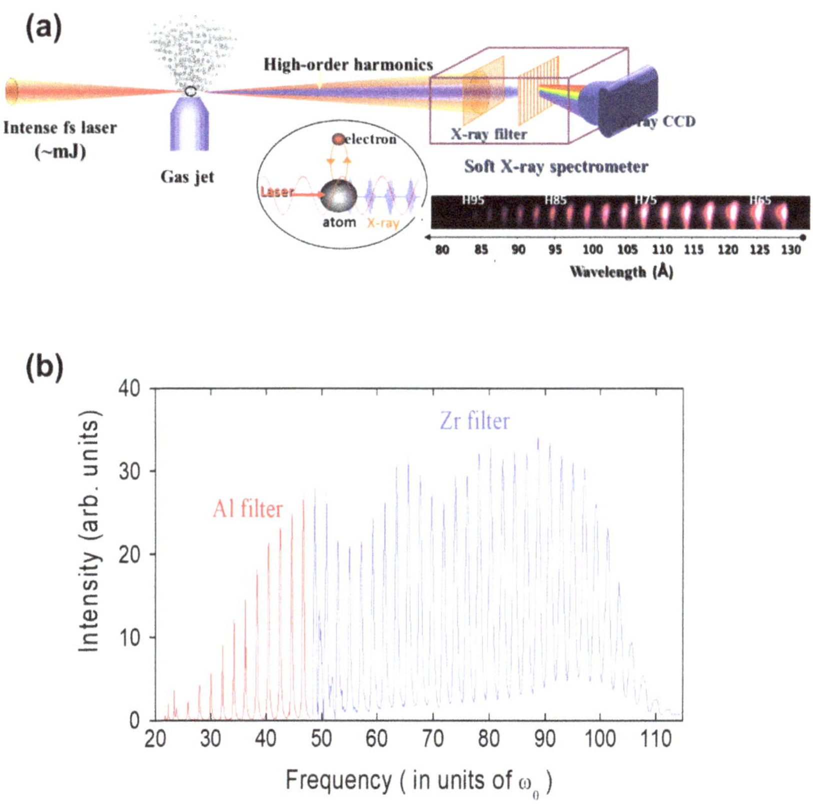


**Figure 5.3.** (a) Schematic layout of HHG experiment and (b) a measured high-order harmonic spectrum. The high harmonic spectrum was taken with a 26 fs 820 nm laser pulse with intensity of $2 \times 10^{16}$ W cm$^{-2}$ and a 1 mm 100 Torr helium gas jet.

for HHG, which typically consists of focusing a femtosecond intense laser pulse on a gaseous medium and measuring the harmonic spectrum with an extreme-ultraviolet (EUV) spectrometer. Because the driving laser pulse and harmonic radiations co-propagate after the plasma medium, a set of thin metal filters is frequently used to block the driving laser pulse and scattered lights. For HHG experiments, the laser intensity is typically in the $10^{13}$–$10^{16}$ W cm$^{-2}$ range, and the target density is in $10^{17}$ – $10^{19}$ cm$^{-3}$ range, with lengths ranging from sub-mm to a few cm [8]. Since HHGs do not necessitate a high laser intensity on target, a few mJ of laser energy is sufficient, which can be operated at a high repetition rate of over kHz. Figure 5.3(b) depicts a typical experimental result of an observed high-order harmonic spectrum measured with two sets of aluminum and zirconium metal filters. Because a tiny hole in a metal filter can destroy the high harmonic signal on the EUV charge-coupled device camera with scattered lights, metal filters are at least used as a pair of thin metal foils of 100s of nanometers. When driving a laser pulse with an 800 nm wavelength, the wavelength of the 101st harmonic is 7.9 nm, which is in the EUV range, whereas driving a laser pulse with a very short pulse duration (10 fs) or a long wavelength (mid-infrared) can produce water-window soft x-rays [3–7]. Furthermore, HHG is a coherent process that generates spatially and temporally coherent short-wavelength radiation [11–13].

The high-order harmonics are separated equally by twice the frequency of the fundamental laser field, which can form a train of attosecond pulses in the time domain if adjacent harmonics have a well-defined phase difference, or are phase-locked, as in the mode-locking laser described in chapter 2. Each attosecond pulse in the pulse train can have a duration of as short as $\tau_L/N$, where $\tau_L$ is the period of one optical cycle of the driving laser field and $N$ is the number of phase-locked harmonics. Attosecond pulses can be produced by perfectly phased-locked five high-order harmonics driven by an 800 nm laser pulse. Figure 5.2(b) depicts an example of an attosecond pulse train formed using a propagation code based on the wave equation and the strong field approximation (SFA) model [14]. Filtering the continuous harmonic spectrum in the cutoff region driven by carrier-envelope-phase-controlled few-cycle laser pulses can result in a single burst of attosecond pulses. The isolated attosecond pulse has significant advantages in ultrafast spectroscopy, allowing for temporal measurements of previously instantaneous photo-ionization processes [15–18].

HHG can be simulated by solving the time-dependent Schrödinger equation (TDSE) [19]. For the case of $U_p \geqslant I_p \gg 1$, the TDSE in length gauge with laser electric field amplitude of $E$, can be expressed in atomic units as

$$i \mid \psi(\boldsymbol{x}, t)\rangle = \left[-\frac{1}{2}\nabla^2 + V(\boldsymbol{x}) - E \cdot \cos(t)x\right] \mid \psi(\boldsymbol{x}, t)\rangle. \tag{5.2}$$

The TDSE calculation is a full quantum simulation on the HHG, which necessitates a significant amount of computational power. A semi-quantum (or semi-classical) simulation with SFA [20] can significantly reduce the computational power required to simulate the macroscopic behavior of HHG from a large number of atoms. The SFA model makes three assumptions, which are as follows:

(i) Neglecting all bound states except ground state
(ii) Neglecting the depletion of the ground state
(iii) Atomic potential has no effect on the electron in the continuum state.

The second assumption is valid when the laser intensity is lower than the saturation intensity ($U_p < U_{\text{sat}}$). Under these conditions, the time-dependent wave function can be expressed as

$$|\,\psi(t)\rangle = e^{iI_p t}\left(a(t)|0\rangle + \int \boldsymbol{b}(v,\, t)|v\rangle d^3 v\right), \tag{5.3}$$

where $a(t)$ and $\boldsymbol{b}(v,\, t)$ are, respectively, the amplitudes of the ground and continuum states. The assumption (ii) permits $a(t)$ to be set equal to 1. The Schrödinger equation (equation (5.2)) can then be written with the wave function (equation (5.3)) by

$$\dot{b}(v,\, t) = -i\left(\frac{v^2}{2} + I_p\right) b(v,\, t) - E\cos(t)\frac{\partial b(v,\, t)}{\partial v_x} + iE\cos(t) d_x(v) \tag{5.4}$$

where the $d(v) = \langle v \mid \boldsymbol{x} \mid 0\rangle$ is the element of the dipole matrix for bound-free transition and $d_x(v)$ is the component parallel to the polarization axis. Equation (5.4) can be solved as follows:

$$\boldsymbol{b}(v,\, t) = i\int_0^t dt'\left[E\cos(t')d_x(v + A(t) - A(t'))\cdot e^{-i\int_{t'}^{t}\left[v + A(t) - A(t'')^2/2 + I_p\right]dt''}\right] \tag{5.5}$$

where $A(t) = (-E\sin(t),\, 0,\, 0)$ is the vector potential of the laser field. The expression for the time-dependent dipole momentum ($\langle\psi(t)|\; x \mid \psi(t)\rangle$) with canonical momentum ($\boldsymbol{p} = v + A(t)$) is

$$x(t) = i\int_0^t dt' \int d^3\boldsymbol{p}\quad [E\cos(t')d_x(\boldsymbol{p} - A(t'))\cdot d_x^*(\boldsymbol{p} - A(t))e^{-iS(\boldsymbol{p},\, t,\, t')}] + c.\, c. \tag{5.6}$$

where

$$S\,(\boldsymbol{p},\, t,\, t') = \int_{t'}^{t} dt''\left[\frac{(\boldsymbol{p} - A(t''))^2}{2} + I_p\right]. \tag{5.7}$$

In equation (5.6), $d_x(\boldsymbol{p} - A(t'))$ represents the probability of transition from bound state to continuum state at time $t'$, $d_x^*(\boldsymbol{p} - A(t))$ is for the returning transition at $t$, and $S(\boldsymbol{p},\, t,\, t')$ denotes the quasi-classical action according to free motion in period between $t'$ and $t$ in laser field with momentum $\boldsymbol{p}$. The major contribution to the integral over $\boldsymbol{p}$ in equation (5.6) can be approximated by the saddle-point method, which uses stationary points of the classical action given by

$$\nabla_p S(\boldsymbol{p},\, t,\, t') = \boldsymbol{x}(t) - \boldsymbol{x}(t') = 0. \tag{5.8}$$

Thus, the integral over $\boldsymbol{p}$ in equation (5.6) leads to

$$x(t) = i\int_0^{\infty} d\tau \left[ \left( \pi / \left( \epsilon + \frac{i\tau}{2} \right) \right)^{\frac{3}{2}} d_x^*(p_{st}(t, \tau) - A_x(t)) \cdot d_x(p_{st}(t, \tau) - A_x(t - \tau)) \cdot E\cos(t - \tau) \cdot e^{-iS_{st}(t, \tau)} \right] + c.\, c.. \tag{5.9}$$

where $\tau = t - t'$ is returning time that is the period between ionization and recombination. Momentum and classical action can be expressed at a stationary point as

$$p_{st}(t, \tau) = E[\cos(t) - \cos(t - \tau)]/\tau, \tag{5.10}$$

$$S_{st}(t, \tau) = \frac{1}{2}\int_{t-\tau}^{t} dt''[(p_{st} - A(t'')^2] = (I_p + U_p)\tau - 2U_p[1 - \cos(\tau)]/\tau - U_pC(\tau) \cdot \cos(2t - \tau) \tag{5.11}$$

where the $C(\tau)$ is defined by

$$C(\tau) = \sin(\tau) - 4\sin^2(\tau/2)/\tau. \tag{5.12}$$

The $\epsilon$, an infinitesimal parameter derived from the regularized Gaussian integration over $\boldsymbol{p}$ around the saddle point, is associated with quantum diffusion and cutoff contribution from multicycle returning. The equation (5.9) can be easily calculated by numerical integration over $\tau$, which provides the semi-classical response of a single atom in a strong laser field for HHG.

There are three distinct types of classical electron trajectories for the generation of high-order harmonics as depicted in figure 5.4: short, long, and cutoff trajectories [14, 21–24]. The cutoff trajectory provides the highest kinetic, whereas the short and

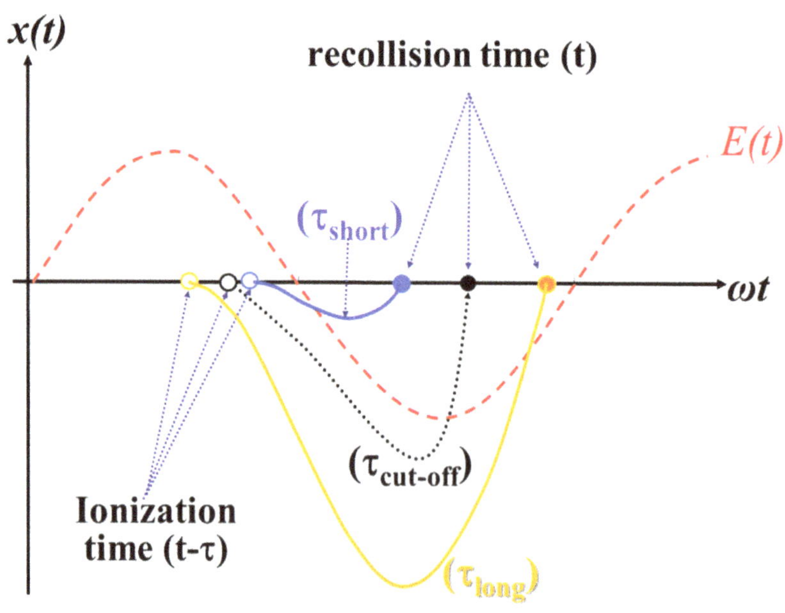


**Figure 5.4.** An illustration of different electron trajectories: short (blue), long (red), and cutoff (black dotted) trajectories.

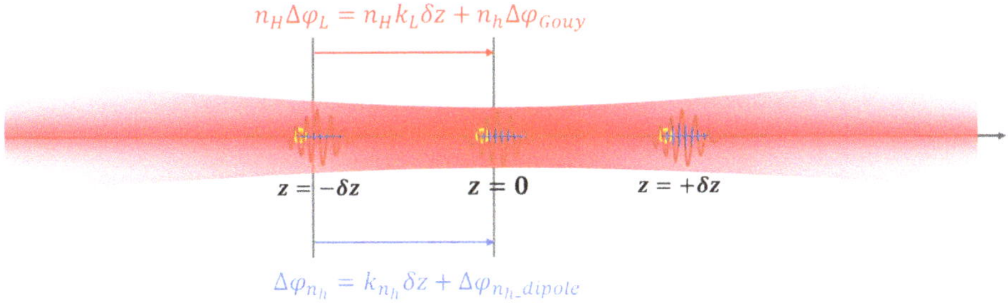


**Figure 5.5.** Conceptual drawing for the phase-matching of HHG.

long trajectories produce nearly the same photon energy with different return times ($\tau$) within one half of the laser cycle. Even though the long and short trajectories that form harmonics in plateau have similar photon energies, the difference in dipole phase, given by equation (5.11), distinguishes the characteristics of the two trajectories. In the phase-matching condition and chirp structure of the high-harmonic spectrum, the dipole phase plays a crucial role. Compared to long trajectories, short trajectories have advantages in phase-matching [21–23] due to their insensitive dependence on laser field variation as a result of their short travel time in the laser field (figure 5.4).

The optimization of the high-order harmonic signal can be achieved through the coherent build-up process during its propagation, which is maintained by ensuring a phase-matching condition (figure 5.5). The phase-matching of the HHG for $n_h{}^{\text{th}}$ harmonic can be accommodated by equalizing the $n_h$ time of laser phase difference and the harmonic phase as the laser propagates; consequently, the harmonics are generated in phase during the laser propagation. Maxwell equations for laser and harmonic pulses with a single atom response calculation by SFA model can be simulated for understanding the macroscopic behavior of HHG, while a simple model can be used to anticipate feasible optimum conditions for HHG. The on-axial phase variation in propagation can be primarily attributed to dispersions and the Gouy phase shift [25] for the driving laser, whereas the harmonic phase can be influenced by dispersion and dipole phase.

To generate bright high-order harmonics, a loose focusing geometry that minimizes Gouy phase shift and dipole phase variation in propagation is advantageous. In the loose focusing geometry, the dispersion difference between the laser pulse and the high-harmonic pulse contributes most to the longitudinal phase difference. In this condition, the on-axial high-harmonic output is proportional to

$$\begin{aligned}&\left|\int_0^{L_{\text{med}}} dz\left[\rho A_{n_h}(z)\cdot e^{-\left(\frac{L_{\text{med}}-z}{2L_{\text{abs}}}\right)}\cdot e^{i\varphi_{n_h}(z)}\right]\right|^2\\ &\approx \rho^2 A_{n_h}^2\frac{4L_{\text{abs}}^2}{1+4\pi^2(L_{\text{abs}}^2/L_{\text{coh}}^2)}\left[1+e^{-\frac{L_{\text{med}}}{L_{\text{abs}}}}-2\cos\left(\frac{\pi L_{\text{med}}}{L_{\text{coh}}}\right)e^{\left(-\frac{L_{\text{med}}}{2L_{\text{abs}}}\right)}\right],\end{aligned} \tag{5.13}$$

where $A_{n_h}(z)$ is the amplitude of the atomic response for the harmonic frequency $n_h\omega_L$ ($\omega_L$ is driving laser frequency), $\varphi_{n_h}(z)$ is the phase of the high harmonic, $\rho$ is medium density, $L_{\mathrm{med}}$ is medium length, and $L_{\mathrm{abs}}$ is absorption length [26–28]. The equation in the second row of equation (5.13) is an approximation based on the assumption of a very loose focusing condition in which the atomic response is nearly constant along the propagation and linear dispersion is the most important factor for phase-matching. $L_{\mathrm{coh}} = \pi/\Delta k$ is the coherence length where $\Delta k = k_{n_h} - n_h k_L$ and $k_L$ is the wave number of the driving laser pulse. When the coherence length is significantly longer than the absorption length ($L_{\mathrm{coh}} > 5L_{\mathrm{abs}}$), the generation of high-order harmonics can be maximized; this phenomenon is known as absorption-limited ($L_{\mathrm{med}} > 3L_{\mathrm{abs}}$) HHG. In addition, the significant plasma dispersion caused by the firstly ionized electron makes phase-matching of the high harmonics from a secondly ionized electron extremely difficult, despite the fact that the secondly ionized electron can produce higher-order harmonics [4]. Thus, it is commonly thought that high-order harmonics are generated by the electrons during the first ionization of atoms.

The photon energy and conversion efficiency are dependent upon the laser intensity and ionization potential of the target atoms. Target atoms with a higher ionization potential, such as helium and neon, can produce higher orders of harmonics, while the saturation intensity can limit the highest achievable order. Few-cycle laser pulses can produce much higher-order harmonics by providing recombining electrons with higher kinetic energy prior to the depletion of ground-state atoms, as the saturation intensity rises with decreasing pulse duration. Atoms with a higher ionization potential can generate more photon energy, but the efficiency of harmonics is lower compared to low-Z atoms, such as Ar, Kr, and Xe, due to the small cross-section of recombination. Consequently, there is a trade-off between photon energy and photon flux of high-order harmonics due to the fact that atoms with a higher ionization potential have a lower recombination rate. HHG can provide an EUV/soft x-ray source with excellent coherence and extremely short pulse duration [29], which has been utilized in various applications including ultrafast spectroscopy [30, 31], coherent imaging [32–34], high-resolution interferometry [11, 35], nonlinear EUV optics [36–38], etc. In addition, the HHG process can provide information on the orbital structures of molecules due to the coherent scattering between the returning wave packet and the ionic core of the molecules [39, 40]. Therefore, high-order harmonics in a non-relativistic regime from a gaseous medium are a useful short-wavelength source with the unique advantages of coherence and short-pulse duration that can be facilitated by compact high-repetition femtosecond lasers, despite the fact that the photon flux per shot [41–43] is not as high as that of x-ray lasers or relativistic high-order harmonics.

## 5.2 X-ray lasers based on laser plasma

X-ray lasers based on laser–plasma interactions make use of the population inversion of highly charged ions that transition in the EUV/soft x-ray regime [44]. Short-wavelength lasers have been anticipated since the beginning of laser development in the 1960s. In 1985, the Lawrence Livermore National Laboratory (LLNL)

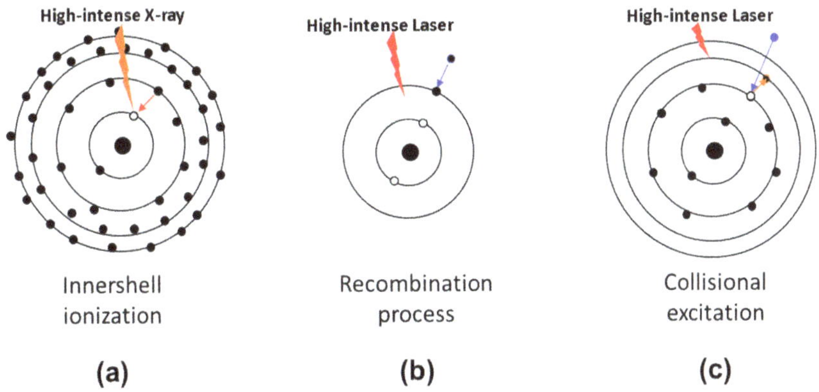


**Figure 5.6.** Schemes for x-ray lasers: inner-shell ionization, recombination, and collisional excitation.

and Princeton University reported the first demonstrations of x-ray lasers [45, 46]. LLNL demonstrated an x-ray laser using a collisional pumping scheme with neon-like selenium ions at 20 nm wavelength, which were pumped by the two beam lines of the laser fusion driver (NOVA) [45]. Using a kJ $CO_2$ laser, a group in Princeton University led by S Suckewer developed a recombination x-ray laser with carbon at approximately 18 nm wavelength [46]. These demonstrations of x-ray lasers in the 1980s required enormous laser energy over kJ per pulse and a ns pulse duration that could not be operated repeatedly.

There are three ways to obtain x-ray lasers with laser–plasma interactions, as depicted in figure 5.6: the inner-shell ionization scheme, the recombination scheme, and the collisional pumping scheme. The inner-shell ionization scheme makes use of the population inversion induced by inner-shell ionization by irradiating atoms with intense x-rays. The depletion of K-shell electrons by inner-shell ionization, for instance, can typically result in a population inversion between K-shell and L-shell transitions. Despite the simplicity of the inner-shell ionization method, the x-ray flash required for population inversion by inner-shell ionization is extremely demanding, and the scheme has been demonstrated using x-ray free-electron laser [47]. The recombination scheme utilizes the population inversion caused by the transition of fully ionized atoms from continuum to upper level [44, 46, 48]. To recombine the electron to a higher level, a low plasma temperature is required, whereas the formation of fully ionized ions in a laser field increases the plasma temperature intrinsically. Therefore, the difficult task of achieving recombination x-ray lasers requires the instantaneous cooling of plasma temperature.

The most successful x-ray laser scheme is collisional excitation, which is based on the collisional excitation of highly ionized ions by energetic electrons over 300 eV. By means of collisions, the energetic electrons excite the ions, while the rapid radiative decay at lower levels creates the conditions for population inversion (figure 5.7). The collisional excitation scheme was demonstrated by transverse pumping geometry, incidence of the laser pulse normal to the surface of the target,

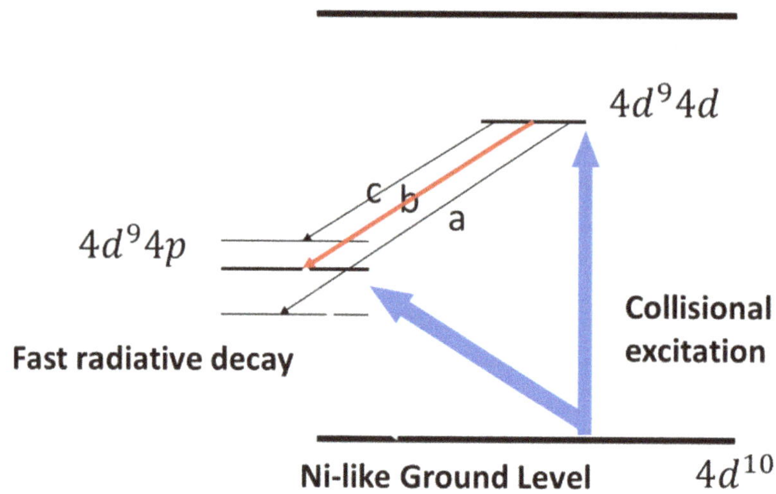


**Figure 5.7.** Schematic energy diagram for collisional excitation scheme in Ni-like ion.

while this geometry required laser pulses with kJ energy and ns pulse duration to obtain the correct amplification of the x-ray lasers. Reducing the pumping laser energy required for developing repetitive x-ray laser pulses has been a significant focus. As ultrashort [45, 49] intense laser systems were developed in the 1990s, the required pumping energy for x-ray lasers was rapidly reduced through the use of a transient pumping scheme by grazing incidence pumping (GRIP) geometry [50–54] and optical-field ionization (OFI) process [55, 56] (figure 5.8). Transient pumping schemes have enabled saturated x-ray lasing with only 1 J of pumping energy, making significant progress over the last two decades.

By employing GRIP geometry and the OFI process, it is possible to achieve transient collisional excitation with compact ultra-intense lasers in an effective manner. The GRIP scheme typically employs two pumping lasers [50]: a long pulse (sub-ns ns pulse duration) is line-focused to a target with normal incidence for generating plasma, and a few-ps short pulse laser with a grazing incidence angle increases the plasma temperature for population inversion. The grazing incidence angle of the main laser pulse can enhance the absorption of the plasma medium and provide traveling wave propagation for the efficient amplification of the x-ray laser. The laser pulse energy is primarily absorbed near the critical density surface when the GRIP scheme employs a solid slab target. Simulations of radiative hydrodynamics can determine the heating process and population inversion dynamics. Numerous x-ray lasers based on Ne-like ions and Ni-like ions have been demonstrated up to the wavelength of approximately 5 nm. In order to simplify the experimental setup, a single-pulse GRIP scheme with a temporally profiled laser pulse was also demonstrated [57, 58]. The OFI x-ray lasers facilitate population inversion using high-energy electrons from the OFI process and intense femtosecond laser pulses with circular polarization [55, 56]. The intense laser pulse with circular

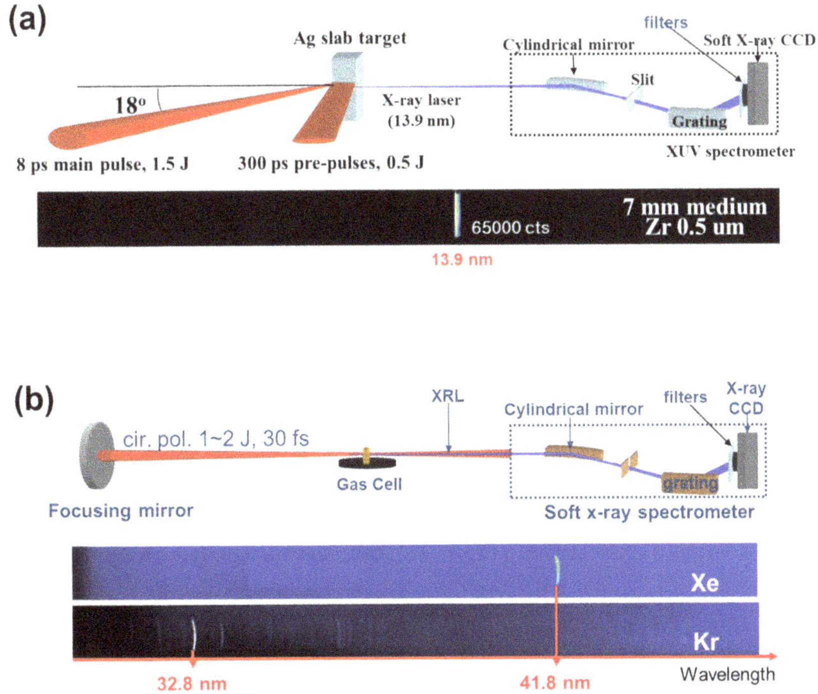


**Figure 5.8.** Experimental layouts and typical x-ray laser spectra of (a) grazing incidence pumping scheme (Ni-like silver x-ray laser at 13.9 nm) and (b) optical-field ionization scheme (Xe laser at 41.8 nm and Kr laser at 32.8 nm).

polarization can initiate the OFI process by supplying electrons with excessive energy over several hundred electron volts, which can cause population inversions. The OFI scheme has demonstrated Xe and Kr x-ray lasers at 41.8 nm and 32.8 nm wavelengths, respectively. Due to the narrow bandwidth of the lasing transition, the pulse duration of these x-ray lasers with transient collisional pumping schemes is in the picosecond range. In addition, the conversion efficiency of x-ray lasers is generally about $10^{-6}$, which provides produce EUV/soft x-ray radiation over μJ energy. Thus the photon flux of x-ray lasers is significantly higher than that of high-order harmonics. These GRIP and OFI x-ray lasers require a pumping laser energy of approximately 1 J, which can be operated at a high repetition rate of 10 Hz or even higher.

Laser–plasma x-ray lasers can generate orders of magnitude more photon flux than HHG below 20 nm wavelength. Due to the absence of a laser cavity, however, the spatial properties and coherence of the beam are inferior to those of high-order harmonics. In the EUV and soft x-ray wavelength range, the high-reflectivity mirrors are very rare due to strong material absorption that makes obstacles to constructing laser cavity for the x-ray lasers. In addition, the relatively short gain lifetime for the transient collisional pumping scheme makes it difficult to construct a suitable laser cavity. Seeding the x-ray laser amplifier with high-order harmonics

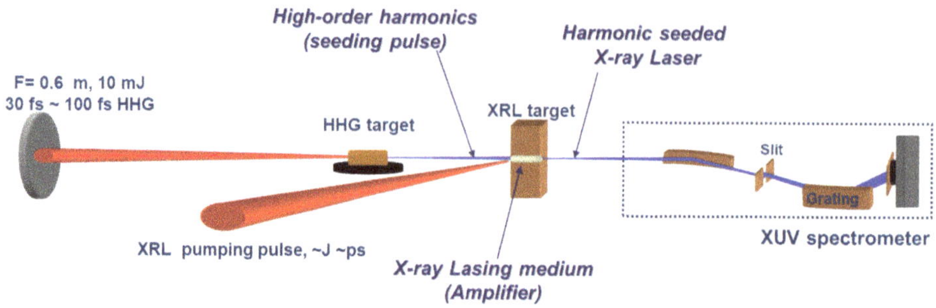


**Figure 5.9.** Experimental layout for harmonic-seeded x-ray laser with a single-pulse grazing incidence pumping scheme.

(figure 5.9) is a method for enhancing the quality of x-ray lasers while maintaining high photon flux [57, 59–62]. The one harmonic in the high-harmonic spectrum must be precisely tuned to the wavelength of the x-ray laser for the harmonic-seed x-ray laser. For OFI and GRIP x-ray lasers, harmonic-seeded x-ray lasers have demonstrated excellent coherence and spatial mode with a pulse energy of about μJ. Recently, a harmonic-seeded Kr x-ray laser with plasma guiding at extremely high plasma density demonstrated that the pulse duration of the harmonic-seeded x-ray laser can be in the femtosecond range [63]. Even though the x-ray free-electron laser is a potent coherent short-pulse x-ray source, harmonic-seeded x-ray lasers can be a complementary coherent short-pulse source in the short-wavelength range that can be generated with a tabletop laser driver.

## 5.3 Betatron radiation

In laser wakefield acceleration (LWFA) process, the injected electron bunch can be oscillated transversely with the centripetal electrostatic force in the plasma bubble when the electron bunches have transverse momentum or are injected off-axis (figure 5.10) [64–68]. In accelerator physics, betatron acceleration refers to the circular acceleration of an electron in a circular electric field induced by a varying magnetic field [69]. In relativistic laser–plasma research, oscillating electrons in the plasma's electric field produce betatron radiation. The electron moves in accordance with the electric field configuration for both betatron accelerators and betatron radiation, although the betatron accelerator employs a circular electric field and betatron radiation relies on a centripetal electron field. Indeed, the term 'betatron' refers to 'beta-particle' which can imply an electron accelerator or a fast electron. Thus, the betatron radiation can be generally interpreted as the radiation emitted by a rapidly oscillating electron.

The electron bunch in the plasma bubble can be simultaneously accelerated and oscillated. For an electron at a transverse position of $\overline{r}_{\perp}$, the restoring force to the oscillation axis is given by Gauss's law as

$$F_{\perp} = -2\pi e^2 n_e \overline{\boldsymbol{r}}_{\perp} = -m_e \omega_p^2 \overline{\boldsymbol{r}}_{\perp}/2. \tag{5.14}$$

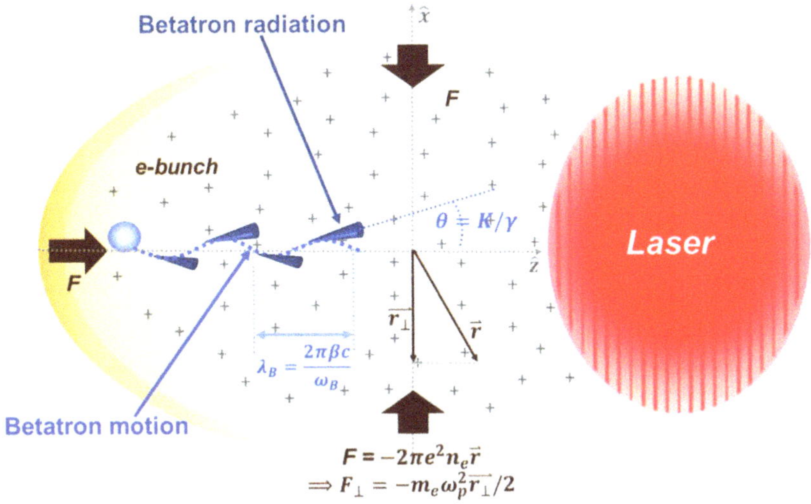


**Figure 5.10.** Concepts of betatron radiation.

Thus, the oscillation motion of electrons can be described as

$$\frac{d\vec{p}}{dt} = -\frac{m_e\omega_p^2\vec{r}_\perp}{2} + m_e c\omega_p\hat{z} \tag{5.15}$$

where $\hat{z}$ is the unit vector along the laser propagation direction. The radiation by the electron motion is given by the Liénard–Wiechert field [70] within the bandwidth of $d\omega$ for center wavelength $\omega$ as follows:

$$\frac{d^2I}{d\omega d\Omega} = \frac{e^2}{16\pi^3\varepsilon_0 c}\cdot\left|\int_{-\infty}^{\infty} e^{i\omega(t-\vec{n}\cdot\vec{r}(t)/c)}\frac{\vec{n}\times[(\vec{n}-\vec{\beta})\times\dot{\vec{\beta}}]}{(1-\vec{\beta}\cdot\vec{n})^2}dt\right|^2 \tag{5.16}$$

where $d\Omega$ is solid angle centered on the observation direction $\vec{n}$. As $\vec{\beta}\cdot\vec{n}$ approaches unity, the radiation power increases, indicating that the relativistic electron motion can produce brilliant betatron radiation. $\vec{n}\times\left[(\vec{n}-\vec{\beta})\times\dot{\vec{\beta}}\right]$ term indicates that the acceleration of electron is required for the radiation and the transverse force to the electron motion is the dominant contributor to radiation. The phase term of the equation (5.16) is related to the Doppler shift, which makes it possible to generate higher-frequency light than the velocity variation frequency of the electrons. The equation (5.16) can be calculated using the numerical solutions of electron trajectories derived from particle-in-cell simulations.

For qualitative understanding of betatron radiation, the equation (5.16) can be approximated by simple harmonic motions based on the force given in equation (5.15). Since only the characteristics of asymptotic on axis will be discussed here, the references should be consulted for the specifics of the derivation [64, 68]. A small betatron oscillation has a wavelength close to

$$\lambda_B = \frac{2\pi}{k_B} \approx \frac{2\pi\sqrt{2\gamma}}{k_p} \tag{5.17}$$

where $k_B$ and $k_p$ are the wave numbers for betatron oscillation and plasma wave, respectively. The betatron radiation is a comparable process to the undulator or wiggler radiation produced by conventional accelerators. The formula for the electron oscillation strength parameter $K$ in the plasma wakefield is:

$$K = \gamma\theta = \gamma k_B r_o \approx 1.33 \times 10^{-10} \cdot \sqrt{\gamma n_e[\mathrm{cm}^{-3}]} \cdot r_0[\mu\mathrm{m}] \tag{5.18}$$

where $r_0$ is the amplitude of the betatron oscillation. For $K \ll 1$, the electron oscillation near the axis is small, and the ion cavity functions as an undulator. For small oscillations close to the axis, the radiation wavelength and frequency can be approximated as

$$\lambda_\beta \approx \lambda_B/(2\gamma^2) = \frac{\sqrt{2}\,\pi c}{\omega_p \gamma^{2/3}},\quad \omega_\beta \approx \sqrt{2}\,\omega_p \gamma^{2/3}. \tag{5.19}$$

As the amplitude of betatron oscillation increases ($K \gg 1$), the radiation spectrum broadens resembling the spectrum of synchrotron radiation, and the plasma bubble assumes the role of a wiggler. The following function characterizes the spectrum of the betatron radiation for large oscillations (figure 5.11) [71]:

$$S(\omega/\omega_c) = (\omega/\omega_c)\int_{(\omega/\omega_c)}^{\infty} K_{5/3}(x)dx \tag{5.20}$$

where the $K_{5/3}$ is the modified Bessel function of second kind with the index of 3/5. The critical frequency ($\omega_c$), is given by

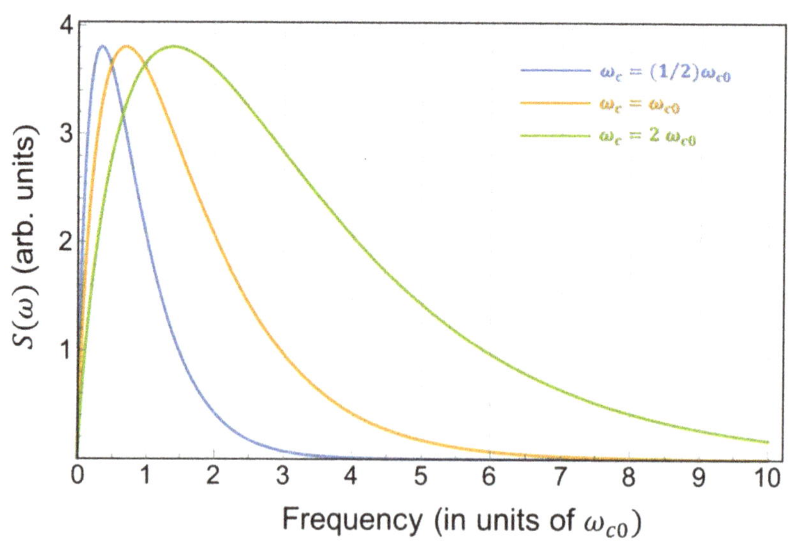


**Figure 5.11.** Typical spectral profile of betatron radiation.

$$\omega_c = (3/2)\gamma^3 cR = (3/2)\gamma^3 c \quad r_0 \quad k_B^2 \approx \frac{3}{8}\frac{\gamma^2 r_0 \omega_p^2}{\pi^2 c} \tag{5.21}$$
$$\hbar\omega_c[\text{keV}] \approx 5 \times 10^{-24}\gamma^2 n_e[\text{cm}^{-3}] \cdot r_0[\mu\text{m}]$$

where $R = 1/(r_0 k_B^2)$ is the effective radius of curvature of the electron trajectory. The average number of photons [71] with mean energy of $\hbar\omega_c$ radiated by an electron is about

$$\langle N_{\hbar\omega_c}\rangle \approx \frac{2\pi}{9}\frac{e^2}{\hbar c} N_0 K \approx 5.6 \times 10^{-3} N_\beta K \tag{5.22}$$

where the $N_\beta$ is the number of betatron oscillation, which can be calculated using the ratio of medium length to betatron wavelength ($N_\beta = L_{med}/\lambda_B$). For under-dense plasma, thus, the total photon number of betatron radiation for the mean energy corresponding to the critical frequency can be approximated as

$$N_{\beta\_\text{photon}} = \langle N_{\hbar\omega_c}\rangle N_e \approx 1.6 \times r_0[\mu\text{m}] \cdot L[\text{cm}] \cdot n_e[10^{19}\text{cm}^{-3}] \cdot N_e \tag{5.23}$$

where $N_e$ is the number of electrons in the betatron motion. As indicated by equations (5.21) and (5.23), the betatron radiation energy and flux increase with increasing electron density, whereas the LWFA requires a lower electron density for higher electron energy. Therefore, the electron density for betatron radiation should be optimized to make a balance between sufficient electron acceleration and robust betatron motion in order to produce betatron radiation with high energy and flux. The betatron radiation can provide hard x-ray short-pulse radiation with a few μm source size and spectral shape similar to that of a synchrotron, making it an excellent compact x-ray source [72–77] for high-resolution computed tomography, ultrafast x-ray absorption spectroscopy, etc [78, 79].

## 5.4 Other radiation sources

One of the conventional methods for producing an x-ray source from laser–plasma interaction is the production of characteristic x-rays by atomic inner-shell transitions. When an intense laser irradiates an over-dense plasma, the plasma can become sufficiently heated to generate hot electrons capable of ionizing the inner-shell electrons of atoms. After inner-shell ionization, an empty energy level, also known as a core hole, is left in the atom, and the transition between outer shell and inner shell emits quantized photons proportional to the energy difference between the states [80, 81]. The photon energy of transition radiation is specific to the species of atoms and transition states; consequently, they are referred to as 'characteristic x-rays.' In the reference [82], the photon energies of the characteristic x-rays for atoms are listed in detail. Figure 5.11(a) and (b) show the diagram for atomic transition for $K_\alpha(n = 2 \rightarrow 1)$ and $K_\beta(n = 3 \rightarrow 1)$, which are the most frequently applied transitions, and the typical spectral shape of the copper $K$ radiation (figures 5.12 and 5.13).

The bremsstrahlung radiation, which derives from the German words 'bremsen' (to brake) and 'Strahlung' (radiation), is the other well-known x-ray radiation produced by laser–plasma interactions. When hot electrons are deflected by an

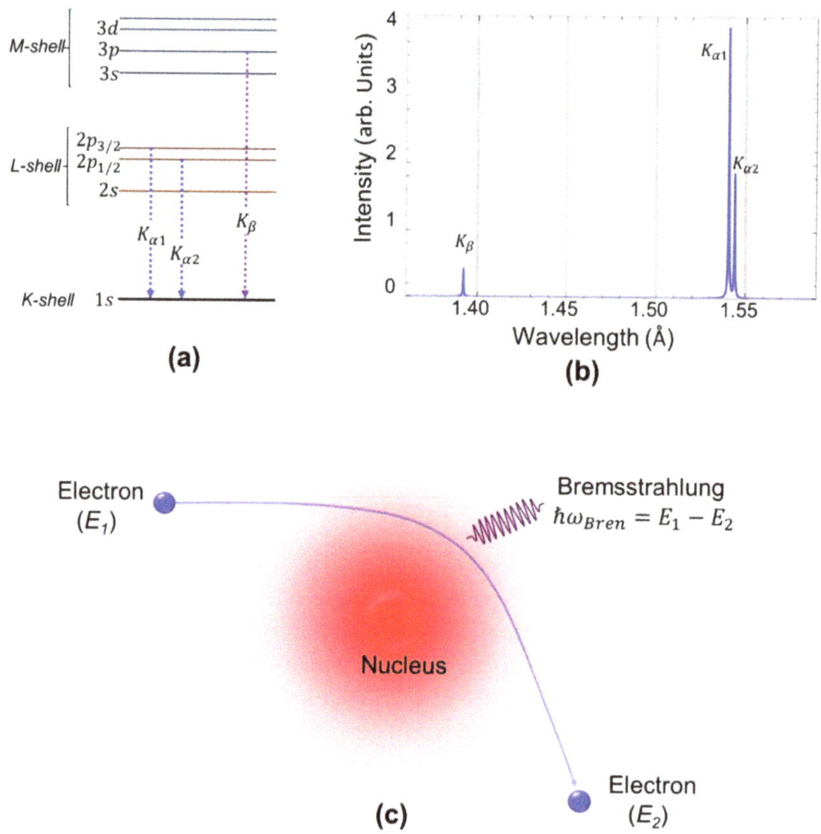


**Figure 5.12.** (a) Energy diagram for characteristic x-ray emission, (b) an exemplary spectral shape of characteristic x-rays from Cu, (c) schematic drawing for bremsstrahlung radiation.

atomic nucleus, bremsstrahlung radiation is emitted (figure 5.12 (c)) [83]. According to the law of energy conservation, the photon energy of bremsstrahlung radiation corresponds to the energy loss of the fast electrons in the atomic potential. Bremsstrahlung radiation has a spectrum that is similar to that of synchrotron radiation. The photon energy and flux of bremsstrahlung radiation are dependent on the electron temperature and atomic potential strength, so efficient plasma heating and a high-Z target are required to produce high-energy, high-flux bremsstrahlung radiation. Moreover, the high-energy electron beam by LWFA can generate extremely high-energy bremsstrahlung radiation by irradiating high-Z materials with the accelerated electron beam. Since the maximum photon energy of bremsstrahlung radiation can be comparable to the incidence electron beam energy, bremsstrahlung radiation with laser-accelerated electron beams can produce gamma rays in the MeV–GeV range [84–86].

As described in section 5.1, the generation of high-order harmonics can be accomplished by the highly nonlinear motion of electrons in an under-dense plasma irradiated by an intense laser pulse in the non-perturbative regime, but not in the

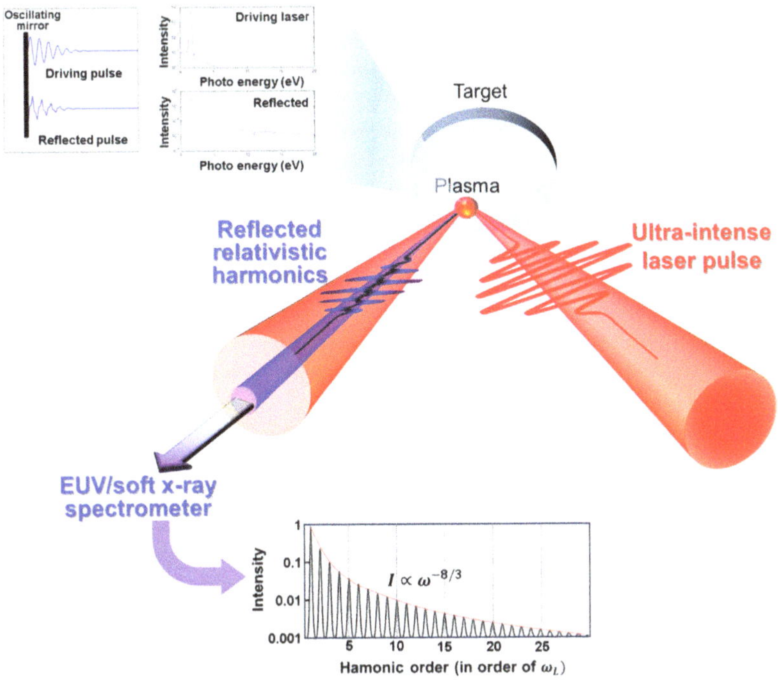


**Figure 5.13.** Relativistic HHG scheme. Upper insert (courtesy of Yong Jin Kwon and Kyung Taec Kim in CoReLS, IBS) shows the relativistic HHG process and lower insert show the typical spectrum shape.

relativistic regime. When an intense laser pulse in the relativistic regime is irradiated on the surface of a solid target to generate an over-dense plasma, the incident laser can be reflected at the critical density layer oscillating at relativistic velocity. When the relativistic oscillating surface reflects the laser field, the Doppler effect, which induces a frequency shift, occurs periodically for each optical cycle [87]. This relativistically oscillating mirror generates high-order harmonics of the reflected laser pulse, resulting in a broad harmonic spectrum ranging from visible to x-rays. Figure 5.13 and the supplementary animation (RHHG_CoReLS, available to download from https://doi.org/10.1088/978-0-7503-2703-9) shows the relativistic HHG process. In contrast to the high-order harmonics produced by gases in the non-relativistic regime, all orders of harmonics are produced in the relativistic HHG. It was known that the radiation intensity for a given harmonic frequency is approximately proportional to $\omega^{-8/3}$ [87]. Due to the fact that relativistic high-order harmonics are generated at the oscillating critical surface, the harmonics are phase-locked, and an attosecond pulse train can be obtained by filtering the harmonics appropriately [88]. The relativistic HHG can enhance the conversion efficiency at least an order of magnitude comparing to the non-relativistic HHG in gas [89]. Therefore, the relativistic high-order harmonics can be used to generate intense attosecond pulses for a variety of applications, including nonlinear x-ray optics. There are also alternative schemes for the generation of coherent x-rays using

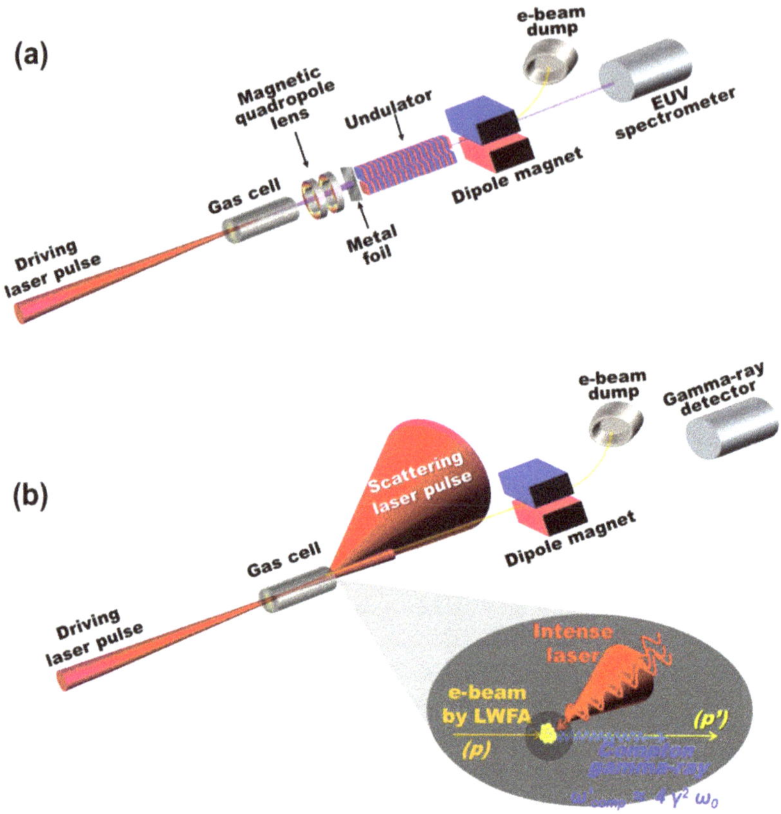


**Figure 5.14.** Method to generate x-rays and gamma rays using laser-accelerated electron beam by (a) magnetic undulator and (b) inverse Compton scattering.

relativistic electron motions, such as the flying mirror [90, 91] and the high harmonic by oscillating electron spike at the edge of a plasma bubble in LWFA [92].

X-ray and gamma-ray sources can be generated using conventional techniques, such as undulator and Compton back-scattering, in conjunction with high-energy electrons produced by LWFA. Recent advances in LWFA have enabled the realization of a highly stable mono-energetic beam suitable for converting these high-energy electron beams into short-wavelength radiations. Figure 5.14 depicts two examples of LWFA-based x-ray and gamma-ray generation methods. The undulator radiation is easily able to generate x-rays with a well-collimated electron beam from LWFA [93], whereas the x-ray free-electron laser necessitates a high-quality and stable electron beam. LWFA recently demonstrated a free-electron laser in the EUV regime based on a sub-GeV electron beam [94, 95]. This LWFA-based free-electron laser demonstrates the potential for downsizing x-ray free-electron lasers, which are typically produced in kilometer-sized conventional accelerator facilities, to tabletop devices for various applications of coherent short-pulse x-ray sources.

Compton back-scattering, also known as inverse Compton scattering, is a technique for producing gamma-ray radiation. The inverse Compton scattering based on the electron beam by LWFA was first demonstrated by reflecting the driving laser pulse at the surface of a thin foil [96]. In the linear Compton regime, the scattered light from a high-energy electron beam can have a maximum photon energy of $E_{\text{Comp}} \approx 4\gamma^2\hbar\omega_L$ for head-on colliding. When an electron beam with an energy of 1 GeV collides with a laser pulse, the maximum photon energy is approximately 24 MeV, which is in the gamma-ray range. Since the Compton scattering provides a defined photon energy for a given scattering angle, the linear inverse Compton scattering can be a method to generate a mono-energetic gamma-ray beam. When the normalized vector potential of the scattering laser pulse is significantly higher than one ($a_0 \gg 1$), the inverse Compton scattering becomes nonlinear, allowing the simultaneous scattering of multiple photons by an electron, and the harmonics of the Compton radiation are produced. This nonlinear Compton scattering extends the radiation spectrum to higher photon energies $E_{\text{Comp}} \approx n \cdot 4\gamma^2\hbar\omega_L$($n$ is harmonic order), while the bandwidth of the radiation becomes broader [97]. The nonlinear Compton scattering experiments with PW laser pulses can be a pathway to explore quantum radiation reaction effect and nonlinear quantum electrodynamics [98–100].

## Practices

1. Calculate cutoff energy and harmonic order of high-order harmonics when hydrogen atoms are irradiated by an 800 nm laser with an intensity of $5 \times 10^{14}\,\text{cm}^{-2}$. In this case, what is the excursion time of the cutoff trajectory?
2. Calculate saturation intensity of a hydrogen atom for 800 nm laser pulse with a pulse duration of 50 fs. In this condition, what are the cutoff energy and harmonic order of high-order harmonics with the saturation intensity?
3. Calculate the minimum pulse duration of one burst in high harmonic pulse train composed the phase-locked harmonics from 81st to 99th orders.
4. Explain three typical schemes of x-ray lasers based on laser–plasma interactions.
5. Which harmonic order can be seeded for Kr OFI x-ray laser for 800 nm driving laser pulse?
6. Calculate the critical energy of betatron radiation for a 100 MeV electron bunch oscillating with amplitude of 5 μm in a plasma bubble formed with an electron density of $10^{18}\ \text{cm}^{-3}$.
7. With the conditions of practice 6, calculate the total photon number of betatron radiation with medium length of 1 cm and bunch charge of 100 pC.
8. Draw spectral profile of the betatron radiation for the conditions of practice 6.
9. What are the photon energy and wavelength of $K_\alpha$ and $K_\beta$ characteristic x-rays from Fe, Co, and Ni?
10. Calculate the shortest wavelength of reflected 800 nm laser pulse by a critical surface oscillating with a maximum velocity of 0.999 99 c.
11. Calculate the maximum photon energy of linear Compton back-scattering between an 800 nm laser pulse and 1 GeV electron beam.

## References


[1] McIntyre I A, Jara H, Johann U, Luk T S, Rhodes C K, McPherson A, Gibson G and Boyer K 1987 Studies of multiphoton production of vacuum-ultraviolet radiation in the rare gases *J. Opt. Soc. Am.* B **4** 595–601

[2] Ferray M, L'Huillier A, Li X F, Lompre L A, Mainfray G and Manus C 1988 Multiple-harmonic conversion of 1064 nm radiation in rare gases *J. Phys. B: At. Mol. Opt. Phys.* **21** L31

[3] Spielmann C, Burnett N H, Sartania S, Koppitsch R, Schnürer M, Kan C, Lenzner M, Wobrauschek P and Krausz F 1997 Generation of coherent x-rays in the water window using 5-femtosecond laser pulses *Science* **278** 661

[4] Gibson E A *et al* 2003 Coherent soft x-ray generation in the water window with quasi-phase matching *Science* **302** 95

[5] Gibson E A, Paul A, Tobey R, Wagner N, Murnane M, Kapteyn H and Christov I P 2004 High-order harmonic generation from argon ions up to 250 EV *OSA Trends Opt. Photon. Ser.* **97** 1093

[6] Takahashi E J and Midorikawa K 2009 Coherent water-window x-ray generation by phase-matched high harmonics in neutral media *Springer Proc. Phys.* **130** 299

[7] Fu Y, Nishimura K, Shao R, Suda A, Midorikawa K, Lan P and Takahashi E J 2020 High efficiency ultrafast water-window harmonic generation for single-shot soft x-ray spectroscopy *Commun. Phys.* **3** 1

[8] Winterfeldt C, Spielmann C and Gerber G 2008 Colloquium: optimal control of high-harmonic generation *Rev. Mod. Phys.* **80** 117

[9] Corkum P B 1993 Plasma perspective on strong field multiphoton ionization *Phys. Rev. Lett.* **71** 1994

[10] Sanpera A, Jönsson P, Watson J B and Burnett K 1995 Harmonic generation beyond the saturation intensity in helium *Phys. Rev.* A **51** 3148

[11] Salières P, Le Déroff L, Auguste T, Monot P, D'oliveira P, Campo D, Hergott J F, Merdji H and Carré B 1999 Frequency-domain interferometry in the XUV with high-order harmonics *Phys. Rev. Lett.* **83** 5483

[12] Aziz N A A, Tarmizi E Z M and Mansor Y 2021 Coherence measurement at different spatial points across high harmonic generation source using different spatial filter *Macromol. Symp.* **397** 2000292

[13] Abdul Aziz N A, Tarmizi E Z M and Dinh K B 2022 Analysis of spatial coherence length of high harmonic generated in Argon *Opt. Quantum Electron.* **54** 1

[14] Kim H T, Tosa V and Nam C H 2006 Synchronized generation of bright high-order harmonics using self-guided and chirped femtosecond laser pulses *J. Phys. B: At. Mol. Opt. Phys.* **39** S265

[15] Hentschel M, Kienberger R, Spielmann C, Reider G A, Milosevic N, Brabec T, Corkum P, Heinzmann U, Drescher M and Krausz F 2001 Attosecond metrology *Nature* **414** 509

[16] Itatani J, Quéré F, Yudin G L, Ivanov M Y, Krausz F and Corkum P B 2002 Attosecond streak camera *Phys. Rev. Lett.* **88** 173903

[17] Baltuška A *et al* 2003 Attosecond control of electronic processes by intense light fields *Nature* **421** 611

[18] Sansone G *et al* 2006 Isolated single-cycle attosecond pulses *Science* **314** 443

[19] Kim J H and Nam C H 2002 Plasma-induced frequency chirp of intense femtosecond lasers and its role in shaping high-order harmonic spectral lines *Phys. Rev.* A **65** 033801

[20] Lewenstein M, Balcou P, Ivanov M Y, L'Huillier A and Corkum P B 1994 Theory of high-harmonic generation by low-frequency laser fields *Phys. Rev.* A **49** 2117

[21] Salières P, L'Huillier A and Lewenstein M 1995 Coherence control of high-order harmonics *Phys. Rev. Lett.* **74** 3776

[22] Balcou P, Sali'eres P, L'Huillier A and Lewenstein M 1997 Generalized phase-matching conditions for high harmonics: the role of field-gradient forces *Phys. Rev.* A **55** 3204

[23] Balcou P, Dederichs A S, Gaarde M B and L'Huillier A 1999 Quantum-path analysis and phase matching of high-order harmonic generation and high-order frequency mixing processes in strong laser fields *J. Phys. B: At. Mol. Opt. Phys.* **32** 2973

[24] Zaïr A *et al* 2008 Quantum path interferences in high-order harmonic generation *Phys. Rev. Lett.* **100** 143902

[25] Gouy L G Sur Une Propriété Nouvelle Des Ondes Lumineuses (https://worldcat.org/ko/title/sur-une-propriete-nouvelle-des-ondes-lumineuses/oclc/25595928)

[26] Constant E, Garzella D, Breger P, Mével E, Dorrer C, Le Blanc C, Salin F and Agostini P 1999 Optimizing high harmonic generation in absorbing gases: model and experiment *Phys. Rev. Lett.* **82** 1668

[27] Bellini M, Corsi C and Gambino M C 2001 Neutral depletion and beam defocusing in harmonic generation from strongly ionized media *Phys. Rev.* A **64** 023411

[28] Kazamias S, Douillet D, Weihe F, Valentin C, Rousse A, Sebban S, Grillon G, Augé F, Hulin D and Balcou P 2003 Global optimization of high harmonic generation *Phys. Rev. Lett.* **90** 193901

[29] Ganeev R A 2023 High-order harmonic generation in laser-induced low-density plasma: past and recent achievements *Appl. Phys. B Lasers Opt.* **129** 1

[30] Bruner B D, Narovlansky-Uzan J, Arusi-Parpar T and Marangos J P 2016 Development of high harmonic generation spectroscopy of organic molecules and biomolecules *J. Phys. B: At. Mol. Opt. Phys.* **49** 132001

[31] Geneaux R, Marroux H J B, Guggenmos A, Neumark D M and Leone S R 2019 Transient absorption spectroscopy using high harmonic generation: a review of ultrafast x-ray dynamics in molecules and solids *Philos. Trans. R. Soc.* A **377** 20170463

[32] Sandberg R L *et al* 2007 Lensless diffractive imaging using tabletop coherent high-harmonic soft-X-ray beams *Phys. Rev. Lett.* **99** 098103

[33] Park S B, Singhal H, Nam C H and Lee K H 2013 Ultrafast direct imaging using a single high harmonic burst *Opt. Lett.* **38** 1253–5

[34] Kfir O *et al* 2017 Nanoscale magnetic imaging using circularly polarized high-harmonic radiation *Sci. Adv.* **3** eaao4641

[35] Austin D R, Arrell C A, Walmsley I A, Wyatt A S, Witting T, Tisch J W G, Frank F and Marangos J P 2011 Lateral shearing interferometry of high-harmonic wavefronts *Opt. Lett.* **36** 1746–8

[36] Nabekawa Y, Hasegawa H, Takahashi E J and Midorikawa K 2005 Production of doubly charged helium ions by two-photon absorption of an intense sub-10-fs soft x-ray pulse at 42 EV photon energy *Phys. Rev. Lett.* **94** 043001

[37] Nabekawa Y, Shimizu T, Okino T, Furusawa K, Hasegawa H, Yamanouchi K and Midorikawa K 2006 Interferometric autocorrelation of an attosecond pulse train in the single-cycle regime *Phys. Rev. Lett.* **97** 153904

[38] Takahashi E J, Lan P, Mücke O D, Nabekawa Y and Midorikawa K 2013 Attosecond nonlinear optics using gigawatt-scale isolated attosecond pulses *Nat. Commun.* **4** 1

[39] Itatani J, Levesque J, Zeidler D, Niikura H, Pé Pin H, Kieffer J C, Corkum P B and Villeneuve D M 2004 Tomographic imaging of molecular orbitals. *Nature* **432** 867–71
[40] Yun H, Lee K M, Sung J H, Kim K T, Kim H T and Nam C H 2015 Resolving multiple molecular orbitals using two-dimensional high-harmonic spectroscopy *Phys. Rev. Lett.* **114** 153901
[41] Takahashi E, Nabekawa Y, Otsuka T, Obara M and Midorikawa K 2002 Generation of highly coherent submicrojoule soft x rays by high-order harmonics *Phys. Rev.* A **66** 021802
[42] Kim H T, Kim I J, Lee D G, Hong K H, Lee Y S, Tosa V and Nam C H 2004 Optimization of high-order harmonic brightness in the space and time domains *Phys. Rev. A—At. Mol. Opt. Phys.* **69** 031805
[43] Kim I J, Kim C M, Kim H T, Lee G H, Lee Y S, Park J Y, Cho D J and Nam C H 2005 Highly efficient high-harmonic generation in an orthogonally polarized two-color laser field *Phys. Rev. Lett.* **94** 243901
[44] Suckewer S and Jaeglé P 2009 X-ray laser: past, present, and future *Laser Phys. Lett.* **6** 411
[45] Matthews D L *et al* 1985 Demonstration of a soft X-ray amplifier *Phys. Rev. Lett.* **54** 110
[46] Suckewer D V S, Skinner C H, Milchberg H and Keane C 1985 Suckewer_PRL55_1st_XRL_1985.Pdf *Phys. Rev. Lett.* **55** 1753
[47] Rohringer N *et al* 2012 Atomic inner-shell X-ray laser at 1.46 nanometres pumped by an x-ray free-electron laser *Nature* **481** 488
[48] Midorikawa K, Toyoda K, Nagata Y, Obara M and Kubodera S 1995 An optical field-induced ionization x-ray laser using a preformed plasma scheme *IEEE J. Sel. Top. Quantum Electron.* **1** 931
[49] Daido H, Kato Y, Murai K, Ninomiya S, Kodama R, Yuan G, Oshikane Y, Takagi M, Takabe H and Koike F 1995 Efficient soft x-ray lasing at 6 to 8 Nm with nickel-like Lanthanide ions *Phys. Rev. Lett.* **75** 1074
[50] Nickles P V, Shlyaptsev V N, Kalachnikov M, Schnürer M, Will I and Sandner W 1997 Short pulse x-ray laser at 32.6 nm based on transient gain in ne-like titanium *Phys. Rev. Lett.* **78** 2748
[51] Keenan R, Dunn J, Shlyaptsev V N, Smith R F, Patel P K and Price D F 2003 Efficient pumping schemes for high average brightness collisional x-ray lasers *Soft X-Ray Lasers Appl.* V **5197** 213
[52] Lindau F *et al* 2007 Optimization toward a high-average-brightness soft-x-ray laser pumped at grazing incidence *Opt. Lett.* **32** 139–41
[53] Wang Y, Wang S, Shlyaptsev V N, Rocca J J, Berrill M and Rockwood A 2018 Compact gain-saturated x-ray lasers down to 6.85 Nm and amplification down to 5.85 nm *Optics* **5** 257–62
[54] Wang S *et al* 2019 High-power ultrashort pulse lasers to pump plasma-based soft x-ray lasers *IEEE J. Sel. Top. Quantum Electron.* **25** 1
[55] Sebban S *et al* 2001 Saturated amplification of a collisionally pumped optical-field-ionization soft x-ray laser at 41.8 nm *Phys. Rev. Lett.* **86** 3004
[56] Sebban S *et al* 2002 Demonstration of a Ni-like Kr optical-field-ionization collisional soft x-ray laser at 32.8 nm *Phys. Rev. Lett.* **89** 253901
[57] Kim H T *et al* 2008 Characteristics of a Ni-like silver x-ray laser pumped by a single profiled laser pulse pumping scheme using a single *J. Opt. Soc. Am.* B **25** 76
[58] Kim H *et al* 2008 Demonstration of a saturated Ni-like Ag x-ray laser pumped by a single profiled laser pulse from a 10-Hz Ti:Sapphire laser system *Phys. Rev.* A **77** 023807

[59] Zeltoun P *et al* 2004 A high-intensity highly coherent soft x-ray femtosecond laser seeded by a high harmonic beam *Nature* **431** 426
[60] Wang Y, Granados E, Larotonda M A, Berrill M, Luther B M, Patel D, Menoni C S and Rocca J J 2006 High-brightness injection-seeded soft-x-ray-laser amplifier using a solid target *Phys. Rev. Lett.* **97** 123901
[61] Luther B, Wang Y, Pedaci F, Granados E, Berrill M and Rocca J J 2008 Highly coherent injection-seeded 13.2 nm tabletop soft x-ray laser *Opt. Lett.* **33** 491–3
[62] Depresseux A *et al* 2015 Demonstration of a circularly polarized plasma-based soft-x-ray laser *Phys. Rev. Lett.* **115** 083901
[63] Depresseux A *et al* 2015 Table-top femtosecond soft x-ray laser by collisional ionization gating *Nat. Photon.* **9** 817
[64] Esarey E, Shadwick B A, Catravas P and Leemans W P 2002 Synchrotron radiation from electron beams in plasma-focusing channels *Phys. Rev. E—Stat. Physics, Plasmas, Fluids, Relat. Interdiscip. Top* **65** 15
[65] Rousse A *et al* 2004 Production of a KeV x-ray beam from synchrotron radiation in relativistic laser-plasma interaction *Phys. Rev. Lett.* **93** 135005
[66] Malka V, Er J, Gauduel Y A, Lefebvre E, Rousse A and Phuoc K T A 2008 Principles and applications of compact laser—plasma accelerators *Nat. Phys.* **4** 447
[67] Esarey E, Schroeder C and Leemans W 2009 Physics of laser-driven plasma-based electron accelerators *Rev. Mod. Phys.* **81** 1229
[68] Corde S, Ta Phuoc K, Lambert G, Fitour R, Malka V, Rousse A, Beck A and Lefebvre E 2013 Femtosecond x rays from laser-plasma accelerators *Rev. Mod. Phys.* **85** 1
[69] Kerst D W 1940 Acceleration of electrons by magnetic induction *Phys. Rev.* **58** 841
[70] Jackson J D 1999 *Classical Electrodynamics* 3rd edn (New York: Wiley)
[71] Kostyukov I, Kiselev S and Pukhov A 2003 X-ray generation in an ion channel *Phys. Plasmas* **10** 4818
[72] Albert F *et al* 2017 Observation of betatron x-ray radiation in a self-modulated laser wakefield accelerator driven with picosecond laser pulses *Phys. Rev. Lett.* **118** 134801
[73] Cipiccia S *et al* 2011 Gamma-rays from harmonically resonant betatron oscillations in a plasma wake *Nat. Phys.* **7** 867
[74] Thomas A G R and Krushelnick K 2009 Betatron x-ray generation from electrons accelerated in a plasma cavity in the presence of laser fields *Phys. Plasmas.* **16** 103103
[75] Albert F *et al* 2018 Betatron x-ray radiation from laser-plasma accelerators driven by femtosecond and picosecond laser systems *Phys. Plasmas* **25** 56706
[76] Ferri J *et al* 2018 High-brilliance betatron γ-ray source powered by laser-accelerated electrons *Phys. Rev. Lett.* **120** 254802
[77] Albert F *et al* 2013 Angular dependence of betatron x-ray spectra from a laser-wakefield accelerator *Phys. Rev. Lett.* **111** 235004
[78] Albert F and Thomas A G R 2016 Applications of laser wakefield accelerator-based light sources *Plasma Phys. Control. Fusion* **58** 103001
[79] Albert F 2023 Principles and applications of x-ray light sources driven by laser wakefield acceleration *Phys. Plasmas* **30** 50902
[80] Serbanescu C, Fourmaux S, Kieffer J-C, Kincaid R and Krol A 2009 A K-alpha x-ray source using high energy and high repetition rate laser system for phase contrast imaging *Proc. SPIE—Int. Soc. Opt. Eng.* **7451** 745115

[81] Wang H J, Li Z R and Bin Chen Z 2020 Kα x-ray emission from nanowire Cu targets driven by femtosecond laser pulses for x-ray conversion and backlight imaging *ACS Omega* **5** 20765–72

[82] *X-Ray Elements* (https://xdb.lbl.gov/Section1/Periodic_Table/X-ray_Elements.html).

[83] Armstrong C D *et al* 2019 Bremsstrahlung emission profile from intense laser-solid interactions as a function of laser focal spot size *Plasma Phys. Control. Fusion* **61** 034001

[84] Döpp A *et al* 2016 A bremsstrahlung gamma-ray source based on stable ionization injection of electrons into a laser wakefield accelerator *Nucl. Instrum. Methods Phys. Res. Sect. A Accel. Spectrometers, Detect. Assoc. Equip.* **830** 515

[85] Li Y F, Shaisultanov R, Chen Y Y, Wan F, Hatsagortsyan K Z, Keitel C H and Li J X 2020 Polarized ultrashort brilliant multi-GeV γ rays via single-shot laser-electron interaction *Phys. Rev. Lett.* **124** 014801

[86] Hannasch A *et al* 2021 Compact spectroscopy of KeV to MeV X-rays from a laser wakefield accelerator *Sci. Rep.* **11** 1

[87] Teubner U and Gibbon P 2009 High-order harmonics from laser-irradiated plasma surfaces *Rev. Mod. Phys.* **81** 445

[88] Kormin D, Borot A, Ma G, Dallari W, Bergues B, Aladi M, Földes I B and Veisz L 2018 Spectral interferometry with waveform-dependent relativistic high-order harmonics from plasma surfaces *Nat. Commun.* **9** 1

[89] Dromey B *et al* 2006 High harmonic generation in the relativistic limit *Nat. Phys.* **2** 456

[90] Kando M *et al* 2009 Enhancement of photon number reflected by the relativistic flying mirror *Phys. Rev. Lett.* **103** 235003

[91] Kim I J *et al* 2012 Relativistic frequency upshift to the extreme ultraviolet regime using self-induced oscillatory flying mirrors *Nat. Commun.* **3** 1

[92] Pirozhkov A S *et al* 2014 High order harmonics from relativistic electron spikes *New J. Phys.* **16** 093003

[93] Fuchs M *et al* 2009 Laser-driven soft-x-ray undulator source *Nat. Phys.* **5** 826

[94] Wang W *et al* 2021 Free-electron lasing at 27 nanometres based on a laser wakefield accelerator *Nature* **595** 516

[95] Labat M *et al* 2022 Seeded free-electron laser driven by a compact laser plasma accelerator *Nat. Photon.* **17** 150

[96] Phuoc K T, Corde S, Thaury C, Malka V, Tafzi A, Goddet J P, Shah R C, Sebban S and Rousse A 2012 All-optical compton gamma-ray source *Nat. Photon.* **6** 308–11

[97] Hannasch A, LaBerge M, Zgadzaj R, Cabadağ J P C, Garcia A L, Kurz T, Cowan T, Schramm U, Irman A and Downer M C 2021 Nonlinear inverse compton scattering from a laser wakefield accelerator and plasma mirror arXiv 2107.00139

[98] Mourou G A, Korn G, Sandner W and Collier J L 2011 *ELI—Extreme Light Infrastructure: Science and Technology with Ultra-Intense Lasers* (Whitebook)

[99] Tanaka K A *et al* 2020 Current status and highlights of the ELI-NP research program *Matter Radiat. Extrem.* **5** 024402

[100] Borneis S *et al* 2021 Design, installation and commissioning of the ELI-Beamlines high-power, high-repetition rate HAPLS laser beam transport system to P3 *High Power Laser Sci. Eng.* **9**

**IOP** Publishing

Laser–Plasma Accelerators and Radiation Sources

**Hyung Taek Kim**

# Chapter 6

# Applications of laser–plasma sources

Ultra-intense lasers are a versatile instrument for producing various photon and particle sources. The interaction between ultra-intense lasers and plasma can produce GeV electrons, 100 MeV protons, coherent EUV/soft x-ray sources, and gamma-ray sources, as discussed in previous chapters. Typically, the radiations and particles beam produced by femtosecond ultra-intense lasers have a pulse duration in the femtosecond range and a tiny source size in the micrometer range, which enables ultrafast precision measurements for a variety of materials and chemicals with rapid dynamics. This chapter introduces briefly the potential applications of ultra-intense laser-based radiation and particle sources, as well as the requirements for laser–plasma applications.

## 6.1 Secondary sources from intense lasers and their applications

Laser–plasma interactions can produce secondary photon and particle sources with a wide range of parameters by adjusting the plasma and laser conditions [1–10]. Compared to conventional particle and radiation technologies based on accelerators, the secondary sources created by ultra-intense lasers have unique properties. Conventional sources based on radio-frequency (RF) accelerators [11, 12] have a size ranging from a few hundred meters to a few kilometers, whereas various sources from ultra-intense laser are small enough to fit in a 100 $m^2$ room. However, the average flux and repetition rate of laser–plasma interaction sources are typically lower than those of conventional sources. Thus, the selection and planning of application research necessitate sophisticated considerations of the laser–plasma source characteristics. Recent studies on secondary sources from laser–plasma interaction typically employ femtosecond high-power lasers based on chirped pulse amplification (CPA) technology, which can produce ultrashort radiations and particle beams in the femtosecond and attosecond range. These ultrashort secondary sources [7, 8, 13–17] can be utilized to investigate the ultrafast dynamics of atoms, molecules, and materials.

doi:10.1088/978-0-7503-2703-9ch6

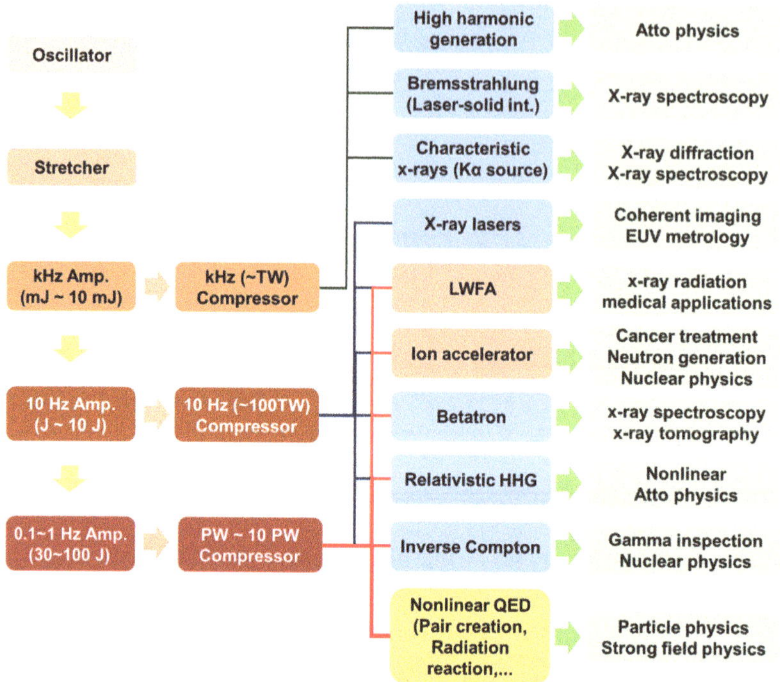


**Figure 6.1.** Various secondary sources based on amplification stages of CPA lasers and their possible application areas.

Another significant benefit of laser–plasma sources is the adaptability of the laser driver. Figure 6.1 depicts the secondary sources that can be generated by CPA femtosecond lasers as well as their potential applications. Each radiation and particle source requires a different accelerator structure in the conventional accelerator source; there are numerous different accelerators for producing different sources and experiments. In contrast, an ultra-intense laser pulse is capable of producing a variety of secondary sources with varying target materials and plasma conditions. Consequently, a laser–plasma radiation and particle source can serve as a testbed for numerous applications involving radiation and particle sources. In addition, the laser pulse can be decoupled in the middle of the amplifier chain and compressed independently to generate sources that do not require the maximum laser system power. Since the amplifying stage of a CPA system with a higher peak power has a lower repetition rate, laser pulses from each amplification stage can be simultaneously distributed to various compressors and target chambers using pulse pickers, typically Pockel's cell. For instance, a petawatt (PW) laser system with a repetition rate of 0.1 Hz [18] can be decoupled from 1 kHz mJ laser pulses and 10 Hz few-J laser pulses, allowing the simultaneous application of three beam lines with different laser specifications to the generation of multiple sources. Therefore, the interaction between ultra-intense lasers and plasmas can produce compact, versatile, and cost-effective sources of radiation and particles.

There are obstacles that must be overcome for various applications, despite the numerous benefits of laser–plasma sources. The operation instability of laser–plasma sources is typically the most critical issue. Plasma is a fluid composed of two charged particle species, ions and electrons, and the majority of plasma dynamics are highly nonlinear due to the hydrodynamic properties of two charged particle fluid [19, 20]. For interactions of ultrashort laser–plasma interactions, hydrodynamic motions do not develop during the main interaction time, but high-energy particle motions in a strong laser field are also complicated phenomena [21, 22]. Moreover, the scale length of laser–plasma interactions is primarily comparable to the plasma wavelength, which is between a few microns and tens of microns. Therefore, the compactness of laser–plasma sources necessitates the use of precise measurement and control techniques in order to stabilize the laser–plasma sources. In spite of these challenges, the extreme-ultraviolet (EUV) source for semiconductor lithography, which employs Sn-specific EUV radiation pumped by $CO_2$ laser pulses [23, 24], is an excellent example of engineering success in laser–plasma application. Consequently, the various laser–plasma sources have a chance of being stabilized and applied to practical applications with significant research resources and engineering efforts.

## 6.2 Applications of particle and radiation sources by relativistic laser–plasma interactions

Radiation and particle sources resulting from relativistic laser–plasma interactions are an emerging technology that has advanced rapidly over the past ten years. There are numerous expectations for the applications of relativistic laser–plasma sources, despite the technological obstacles that must be overcome. Figure 6.2 shows the examples of applications of radiations and particle sources from the relativistic laser-plasma interactions. Ultrafast imaging [25–31] and spectroscopies [32–36] utilizing x-ray lasers and high-order harmonic generation are one of the successful

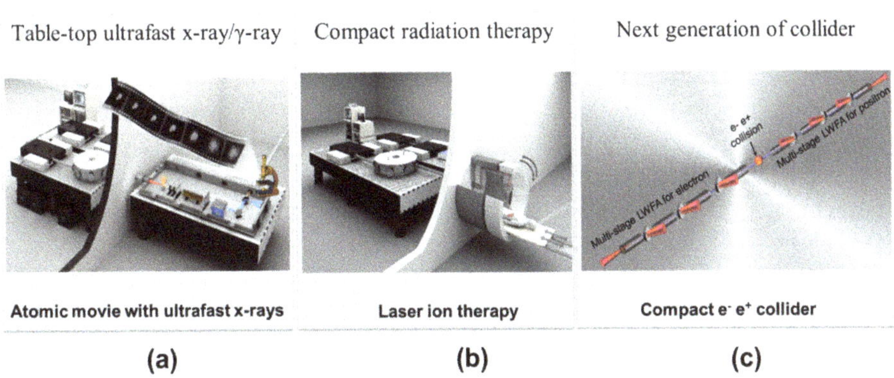


**Figure 6.2.** Examples for the applications of particle and radiation sources from relativistic laser–plasma interactions: (a) electron-positron collider based on LWFA, (b) compact cancer therapy with laser-accelerated ion beams, and (c) ultrafast x-ray microscopy with laser-based tabletop coherent x-ray sources.

applications of non-relativistic laser–plasma sources. Imaging and spectroscopy performance can be enhanced by employing relativistic high-order harmonic generation and laser wakefield acceleration (LWFA)-based short-wavelength sources. The ultrashort x-rays produced by relativistic laser–plasma interactions can enable an ultrafast imaging technique with nanometer-scale spatial resolution and attosecond temporal resolution that can record the movement of atoms during chemical reactions and structural changes in molecules. Moreover, x-rays and gamma rays from relativistic laser plasma can be used for non-destructive high-resolution inspections, nuclear excitation, and high-contrast high-resolution medical images, among other applications.

Important research areas include the applications of relativistic laser–plasma sources to radiation biology and medicine. One of the objectives is the development of cancer treatment technology with laser-accelerated ion and electron beams [37–40]. The cancer therapy with proton and ion sources utilizing cyclotron accelerators is the most advanced radiation therapy due to the localized absorption of the ion energy at the cancer tissue by Bragg peak, which can reduce the side effects of radiation therapy for normal tissue surrounding the cancer. Since each type of cancer tissue can respond differently to different ion species, various ion species' effects must be investigated, but conventional cyclotrons cannot accelerate multiple ion species simultaneously. From the perspective of multiple-ion acceleration, laser–plasma ion accelerations offer unique advantages for studying the biological responses of tissues for different ion species, as the laser ion acceleration mechanism can accelerate multiple ion species in the plasma simultaneously. By using electromagnetic deflection devices, laser plasma's various ion species can be separated and irradiated onto biological samples. In addition, the conventional ion therapy machine's gentry is a complex and expensive device, whereas an optical gentry that delivers an ultra-intense pulse to the patient can be developed for laser ion therapy machines. For laser ion therapy, the maximum energy of ions must be increased by more than twofold, and the dose of the ion beam must be increased by more than one order of magnitude [38, 39]. The alternative method of medical application is the use of laser-accelerated electrons to treat cancer. Even though electron beams do not induce Bragg peaks, high-energy electron beams that are collimated and overlapped can produce a higher local dose than Gamma-knife technology [37, 40]. Therefore, the laser-accelerated ion and electron beam can be candidates for future compact and cost-effective radiation cancer therapy technology.

Development of electron accelerator technology for future electron-positron colliders represents the most difficult application of relativistic laser–plasma interactions [41, 42]. Current RF linear accelerator technologies are insufficient for the construction of electron-positron colliders with center-of-mass energies of several hundred GeV, which are required to explore new regimes of particle physics and produce exotic particles, because the collider requires tens of kilometers of acceleration length. Thus, the plasma accelerators, including LWFA and beam-driven electron accelerations that are also known as plasma wakefield acceleration (PWFA) [41, 43–45], are considered as possible candidates of the future collider technology. For the future collider, LWFA should be revolutionarily improved to have much higher energy, luminosity, and wall-plugged efficiency. PWFA can reduce the size of the accelerator, but not as much as LWFA due to the ultra-short high-energy electron beam driver for PWFA using linear accelerator facilities. Even though the requirements for future colliders are extremely high, pursuing the electron-positron collider technology of the future is an enticing objective that will provide new opportunities for gaining new understandings of the universe. Furthermore, the laser-produced high-energy electron can be used to study nonlinear quantum electrodynamics (QED) such as the Breit–Wheeler process, the trident process, and cascaded gamma-ray generation, among others [7, 46–54]. When electron beams with energies greater than 10 GeV are exposed to an extremely intense laser field with a normalized vector potential greater than 100, the interactions are dominated by the nonlinear QED process. Consequently, the interactions between relativistic lasers and plasmas can pave the way for the discovery of unexplored physical phenomena that can lead to a deeper understanding of the universe.

## References


[1] Umstadter D 2003 Relativistic laser–plasma interactions *J. Phys. D: Appl. Phys.* **36** R151

[2] Gibbon P 2005 *Short Pulse Laser Interactions with Matter: an Introduction* (London: Imperial College Press) 312

[3] Malka V, Er J, Gauduel Y A, Lefebvre E, Rousse A and Phuoc K T A 2008 Principles and applications of compact laser–plasma accelerators *Nat. Phys.* **4** 447

[4] Esarey E, Schroeder C and Leemans W 2009 Physics of laser-driven plasma-based electron accelerators *Rev. Mod. Phys.* **81** 1229

[5] Norreys P A 2009 Laser-driven particle acceleration *Nat. Photonics* **3** 423

[6] Schlenvoigt H-P, Jäckel O, Pfotenhauer S M and Kaluza M C 2010 Laser-based particle acceleration *Advances in Solid State Lasers Development and Applications* ed M Grishin (London: InTech) 565–608

[7] Mourou G A, Korn G, Sandner W and Collier J L 2011 *ELI—Extreme Light Infrastructure: Science and Technology with Ultra-Intense Lasers* (Whitebook)

[8] Corde S, Ta Phuoc K, Lambert G, Fitour R, Malka V, Rousse A, Beck A and Lefebvre E 2013 Femtosecond x rays from laser-plasma accelerators *Rev. Mod. Phys.* **85** 1

[9] Joshi C 2017 Laser-driven plasma accelerators operating in the self-guided, blowout regime *IEEE Trans. Plasma Sci.* **45** 3134

[10] Götzfried J, Döpp A, Gilljohann M F, Foerster F M, Ding H, Schindler S, Schilling G, Buck A, Veisz L and Karsch S 2020 Physics of high-charge electron beams in laser-plasma wakefields *Phys. Rev.* X **10** 041015

[11] Wangler T P 2008 *RF Linear Accelerators* 2nd edn (New York: Wiley)

[12] Wang F, Adolphsen C and Nantista C 2011 Performance limiting effects in X-band accelerators *Phys. Rev. Spec. Top.—Accel. Beams* **14** 1

[13] Dromey B *et al* 2006 High harmonic generation in the relativistic limit *Nat. Phys.* **2** 456

[14] Winterfeldt C, Spielmann C and Gerber G 2008 Colloquium: optimal control of high-harmonic generation *Rev. Mod. Phys.* **80** 117

[15] Depresseux A *et al* 2015 Table-top femtosecond soft x-ray laser by collisional ionization gating *Nat. Photonics* **9** 817

[16] Albert F and Thomas A G R 2016 Applications of laser wakefield accelerator-based light sources *Plasma Phys. Control. Fusion* **58** 103001

[17] Albert F 2023 Principles and applications of x-ray light sources driven by laser wakefield acceleration *Phys. Plasmas* **30** 50902

[18] Sung J H, Lee S K, Lee H W, Yoo J Y and Jeong T M 2017 0.1 Hz 4.0 PW Ti:Sapphire laser at CoReLS *2015 Conference on Lasers and Electro-Optics Pacific Rim* 42 (Optica Publishing Group) 3

[19] Kaw P K 2017 Nonlinear laser–plasma interactions *Rev. Mod. Plasma Phys.* **1** 1

[20] Takabe H and Kuramitsu Y 2021 Recent progress of laboratory astrophysics with intense lasers *High Power Laser Sci. Eng.* **9** e49

[21] Gibbon P and Förster E 1996 Short-pulse laser-plasma interactions *Plasma Phys. Control. Fusion* **38** 769

[22] Salamin Y I, Hu S X, Hatsagortsyan K Z and Keitel C H 2006 Relativistic high-power laser–matter interactions *Phys. Rep.* **427** 41

[23] Yang D-K *et al* 2022 The development of laser-produced plasma EUV light source *Chip* **1** 100019

[24] Khalid A, Wei Y, Rizwan M and Saleem *et al* 2022 High-NA EUV lithography: current status and outlook for the future *Jpn. J. Appl. Phys.* **61** SD0803

[25] Filevich J *et al* 2004 Picosecond-resolution soft-x-ray laser plasma interferometry *Appl. Opt.* **43** 3938

[26] Marchesini S *et al* 2008 Massively parallel x-ray holography *Nat. Photonics* **2** 560

[27] Park J J, Kim D S, Jeon S C, Park J, Lee K H, Lee J H, Kim K N, Yoo J J and Nam C H 2009 Table-top soft x-ray microscope adopting a PMMA phase-reversal zone plate *Opt. InfoBase Conf. Pap.* 34 235

[28] Kim H T *et al* 2011 Single-shot nanometer-scale holographic imaging with laser-driven x-ray laser *Appl. Phys. Lett.* **98** 121105

[29] Kang H C *et al* 2012 Single-pulse coherent diffraction imaging using soft x-ray laser *Opt. Lett.* **37** 1688

[30] Park S B, Singhal H, Nam C H and Lee K H 2013 Ultrafast direct imaging using a single high harmonic burst *Opt. Lett.* **38** 1253–5

[31] Fu Y, Nishimura K, Shao R, Suda A, Midorikawa K, Lan P and Takahashi E J 2020 High efficiency ultrafast water-window harmonic generation for single-shot soft x-ray spectroscopy *Commun. Phys.* **3** 1

[32] Hentschel M, Kienberger R, Spielmann C, Reider G A, Milosevic N, Brabec T, Corkum P, Heinzmann U, Drescher M and Krausz F 2001 Attosecond metrology *Nature* **414** 509

[33] Itatani J, Quéré F, Yudin G L, Ivanov M Y, Krausz F and Corkum P B 2002 Attosecond streak camera *Phys. Rev. Lett.* **88** 173903

[34] Takahashi E J, Lan P, Mücke O D, Nabekawa Y and Midorikawa K 2013 Attosecond nonlinear optics using gigawatt-scale isolated attosecond pulses *Nat. Commun.* **4** 1

[35] Bruner B D, Narovlansky-Uzan J, Arusi-Parpar T and Marangos J P 2016 Development of high harmonic generation spectroscopy of organic molecules and biomolecules *J. Phys. B: At. Mol. Opt. Phys.* **49** 132001

[36] Geneaux R, Marroux H J B, Guggenmos A, Neumark D M and Leone S R 2019 Transient absorption spectroscopy using high harmonic generation: a review of ultrafast x-ray dynamics in molecules and solids *Philos. Trans. R. Soc.* A **377**

[37] Kainz K K, Hogstrom K R, Antolak J A, Almond P R, Bloch C D, Chiu C, Fomytskyi M, Raischel F, Downer M and Tajima T 2004 Dose properties of a laser accelerated electron beam and prospects for clinical application *Med. Phys.* **31** 2053

[38] Daido H, Nishiuchi M and Pirozhkov A S 2012 Review of laser-driven ion sources and their applications *Rep. Prog. Phys.* **75** 056401
[39] Bulanov S V, Wilkens J J, Esirkepov T Z, Korn G, Kraft G, Kraft S D, Molls M and Khoroshkov V S 2014 Laser ion acceleration for hadron therapy *Phys. Usp.* **57** 1149
[40] Kokurewicz K *et al* 2017 *Laser-plasma generated very high energy electrons (VHEEs) in radiotherapy* **10239** 61
[41] Joshi C and Katsouleas T 2003 Plasma accelerators at the energy frontier and on tabletops *Phys. Today* **56** 47
[42] Leemans W and Esarey E 2009 Laser-driven plasma-wave electron accelerators *Phys. Today* **62** 44
[43] Blumenfeld I *et al* 2007 Energy doubling of 42 GeV electrons in a metre-scale plasma wakefield accelerator *Nature* **445** 741
[44] Adli E *et al* 2018 Acceleration of electrons in the plasma wakefield of a proton bunch *Nature* **561** 363
[45] Adli E 2022 Towards a PWFA linear collider—opportunities and challenges *J. Instrum.* **17** T05006
[46] Gonoskov A, Bastrakov S, Efimenko E, Ilderton A, Marklund M, Meyerov I, Muraviev A, Sergeev A, Surmin I and Wallin E 2015 Extended particle-in-cell schemes for physics in ultrastrong laser fields: review and developments *Phys. Rev.* E **92** 023305
[47] Hojbota C I, Kim H T, Kim C M, Pathak V B and Nam C H 2018 Effect of the temporal laser pulse asymmetry on pair production processes during intense laser-electron scattering *Plasma Phys. Control. Fusion* **60** 064004
[48] Vranic M, Klimo O, Korn G and Weber S 2018 Multi-GeV electron-positron beam generation from laser-electron scattering *Sci. Rep.* **8** 1
[49] Hojbota C I, Kim H T, Pathak V B and Nam C H 2019 Influence of polarization on back-reflected E−e+ pair jets from laser-electron collision *Plasma Phys. Control. Fusion* **62** 024003
[50] Gong Z, Hu R H, Yu J Q, Shou Y R, Arefiev A V and Yan X Q 2019 Radiation rebound and quantum splash in electron-laser collisions *Phys. Rev. Accel. Beams* **22** 093401
[51] Geng X S, Ji L L, Shen B F, Feng B, Guo Z, Yu Q, Zhang L G and Xu Z Z 2019 Quantum reflection above the classical radiation-reaction barrier in the quantum electro-dynamics regime *Commun. Phys.* **2**
[52] Tanaka K A *et al* 2020 Current status and highlights of the ELI-NP research program *Matter Radiat. Extrem.* **5** 024402
[53] Torgrimsson G 2020 Nonlinear trident in the high-energy limit: nonlocality, coulomb field, and resummations *Phys. Rev.* D **102** 096008
[54] Keitel C *et al* 2021 Photo-induced pair production and strong field QED on gemini arXiv 2103.06059.

www.ingramcontent.com/pod-product-compliance
Ingram Content Group UK Ltd.
Pitfield, Milton Keynes, MK11 3LW, UK
UKHW051849100926
8556IPUK00005B/541

9 780750 327015